AF877618

Christopher Jung, Jörg Meyer, Achim Streit (Eds.)

Helmholtz Portfolio Theme Large-Scale Data Management and Analysis (LSDMA)

Helmholtz Portfolio Theme Large-Scale Data Management and Analysis (LSDMA)

edited by
Christopher Jung, Jörg Meyer, Achim Streit

KIT Scientific Publishing

Impressum

KIT Scientific Publishing

Karlsruher Institut für Technologie (KIT)
KIT Scientific Publishing
Straße am Forum 2
D-76131 Karlsruhe

KIT Scientific Publishing is a registered trademark
of Karlsruhe Institute of Technology.
Reprint using the book cover is not allowed.

www.ksp.kit.edu

This document – excluding the cover, pictures and graphs – is licensed under a Creative Commons Attribution-Share Alike 4.0 International License (CC BY-SA 4.0): https://creativecommons.org/licenses/by-sa/4.0/deed.en

The cover page is licensed under a Creative Commons Attribution-No Derivatives 4.0 International License (CC BY-ND 4.0): https://creativecommons.org/licenses/by-nd/4.0/deed.en

Print on Demand 2017 – Gedruckt auf FSC-zertifiziertem Papier

ISBN 978-3-7315-0695-9
DOI 10.5445/KSP/1000071931

Contents

Dear Readers,

LSDMA stands for "Large-Scale Data Management and Analysis" and was a portfolio theme funded by the German Helmholtz Association from 2012–2016. Under leadership of the Karlsruhe Institute of Technology (KIT), four Helmholtz centres (KIT, FZ Jülich, DESY, GSI), six universities (University of Hamburg, University of Ulm, Heidelberg University, HTW Berlin, TU Dresden and GU Frankfurt) and the German Climate Computing Centre (DKRZ) joined to enable data-intensive science by optimising data life cycles in selected scientific communities.

Figure 1: The LSDMA Symposium "The Challenge of Big Data in Science" 2016.

In our Data Life Cycle Labs (DLCLs), data experts performed joint R&D together with scientific communities to optimise data management and analysis tools, processes and methods. Complementing the activities in the DLCLs, the Data Services Integration Team (DSIT) focused on the development of generic tools and solutions, which are applied by several scientific communities. Examples are authentication, authorisation, identity management, archiving or metadata.

Overall 78 scientists – among them 21 PhD researchers – were working in LSDMA and have achieved very interesting results ranging from community-specific solutions, e.g. in energy or climate/environmental research, to generic tools and methods, e.g. for meta-data handling or federated AAI. This book gives an overview on these fascinating R&D.

Figure 2: The LSDMA All-Hands Meeting 2016.

In addition, LSDMA organised several annual events: the international symposium on "The Challenge of Big Data in Science", community forums, technical forums and PhD meetings – all these events promoted the enabling of data-intensive science, brought together LSDMA consortium partners with the scientific communities and fostered the spreading and uptake of LSDMA solutions.

New projects originate from the new connections among people in LSDMA and their scientific results, e.g. the DFG-funded MASi project focusses on metadata management for applied sciences[1] and the EC-funded project INDIGO-DataCloud aims at developing a data/compute platform for data-intensive scientific communities provisioned over hybrid einfrastructures[2] Much of the work of LSDMA is meanwhile carried forward in the third round of the Helmholtz programme-oriented funding (PoF-3).

Internationally several LSDMA scientists are actively participating in the Research Data Alliance (RDA) through participating and/or leading working and interest groups as well as severing as elected members in RDA boards such as the Technical Advisory Board (TAB).

I want to express our gratitude to the German Helmholtz Association and the German Federal Ministry of Education and Research for funding the LSDMA portfolio theme.

Have a nice time reading the book.

Dr. Christopher Jung,
Dr. Jörg Meyer,
Prof. Dr. Achim Streit *Lead-PI of the LSDMA Helmholtz Portfolio Theme*

---

[1] https://tu-dresden.de/zih/forschung/projekte/masi
[2] https://www.indigo-datacloud.eu/

Analysis of Environmental Satellite Data with Full Stack of Web Applications

Marek Szuba[a], Parinaz Ameri[a], Jörg Meyer[a]

[a] Karlsruhe Institute of Technology (KIT), Karlsruhe

Abstract We have created a distributed system for storage, processing, three-dimensional visualization and basic analysis of data from Earth-observing satellite experiments such as Envisat MIPAS. The database and the server have been designed for high performance and scalability, whereas the client is highly portable thanks to having been designed as a HTML5- and WebGL-based Web application. The system is based on the so-called MEAN stack, a modern replacement for LAMP, which has steadily been gaining traction among high-performance Web applications. Here, we present an overview of all components of our system.

1 Introduction

Modern remote sensing devices mounted on environmental satellites like MIPAS on Envisat generate enormous amounts of data. From the original raw data various secondary data are derived and disseminated for further analysis. Derived data may differ in geometries and formats leading to a large variety of datasets that need to be harmonized, even though researchers agreed on certain standards and conventions. Traditionally, climate data are stored in file hierarchies on large state-of-the-art storage systems that are financially feasible. For the analysis of data typically detailed expert knowledge is required for specialized programs, e.g. for visualization and knowledge in several high-level programming languages like Java [Ame14].

The progress in modern hardware and the increase of CPU cores allows to change the analysis paradigms towards an on-the-fly parallel pre-processing of data stored as semi-structured data in distributed databases that are provided by RESTful Web services. Researchers just fetch relevant data and perform their analysis including visualization on lightweight clients in close to real-time. In this article we describe the architecture of such a system.
In the next section the components are described and an overview of the system architecture is given. In Section 3 the scalable multi-processing platform `Node Scala` is introduced that is used to perform data pre-processing on a cluster. The client web application `KAGLVis` is described in Section 4, followed by the conclusion.

2 System Components and Architecture

The idea of our system is to index large amounts of heterogeneous data and to be able to query for data instead of reading and parsing files in filesystems. Given the variety of data formats, the high number of data sources, and the volume of the data the requirements for a database management system are the ability for horizontal scaling and a flexible scheme or data model. Horizontal scaling means that data is distributed on several hardware servers, i.e. the capacity of the database system is not limited by the resources of a single hardware server. While changes of a relational database model can be time consuming and cumbersome NoSQL databases do not require a fixed scheme for data. Document-based NoSQL databases fulfill both of our criteria. We implemented our system using a `MongoDB`. Data is stored in semi-structured JSON-documents that consists of key-value pairs allowing for non-scalar values like arrays or nested documents. Horizontal scaling is realized by relaxing guarantees on consistency. However, in our use cases data will be imported once and afterwards read many times. The lack of transactions is no limitation. `MongoDB` provides its own query language that allows to filter and aggregate documents. Complicated queries that include data trans-

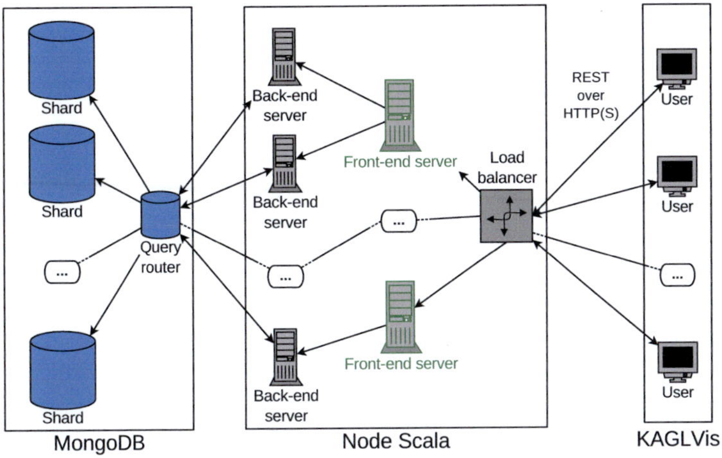

Figure 3: Architecture of our system. For clarity, the diagram omits internal management components of Node Scala (the controllers and the scheduler) as well as assuming only one `MongoDB` query router is in use (and thus hiding the second load balancer).

formations might not be expressible or take too much time. This is why we introduced `Node Scala`, a scalable multi-processing platform. It fetches data from the `MongoDB` and pre-processes the data on a cluster before passing the transformed and reduced data to client application (see Section 3). Our user analysis application is a browser application that runs in all modern web browsers without the need to install further software. The application is called `KAGLVis`. The purpose is to select and visualize environmental satellite data from MIPAS (see Section 4). The complete architecture is depicted in Figure 3.

We have based our system on the so-called MEAN stack, a modern Web-application stack consisting of:

- **M**ongoDB — NoSQL document database
- **E**xpress — Web framework for Node.js
- **A**ngularJS — Web client MVVM framework
- **N**ode.js — event-driven I/O in JavaScript

Using MEAN offers several advantages over more traditional stacks such as LAMP. To begin with, all of its components have been explicitly designed for high performance. Secondly, the fact a single language – JavaScript – is used throughout the stack both simplifies the development environment and simplifies the work of developers. Finally, the common data format of the stack – JSON – maps well to both JavaScript objects and `MongoDB` documents.

3 Node Scala

While undoubtedly optimized for performance, Node.js applications are by design restricted to a single thread. Distributed systems featuring multiple instances of the same application accessed through a common interface such as the same HTTP server, can be constructed using the standard Node.js module Cluster (on a single host) or third-party solutions built on top of it such as StrongLoop Process Manager (which can support multiple hosts), however, neither of these solutions allow for parallel processing. In light of the above, we have designed and developed an alternative solution called Node Scala, which not only allows for distribution of its various components on both a single and multiple hosts but also allows for parallel execution of tasks with intelligent distribution of chunks among its workers [Maa15].
The internal structure of Node Scala is presented in Figure 4. It consists of several component types:

- The **controller** handles start-up and shutdown of other components of the system, even when they run on multiple machines, as well as monitors and restarts them as needed to increase overall robustness. It uses SSH as its command channel.

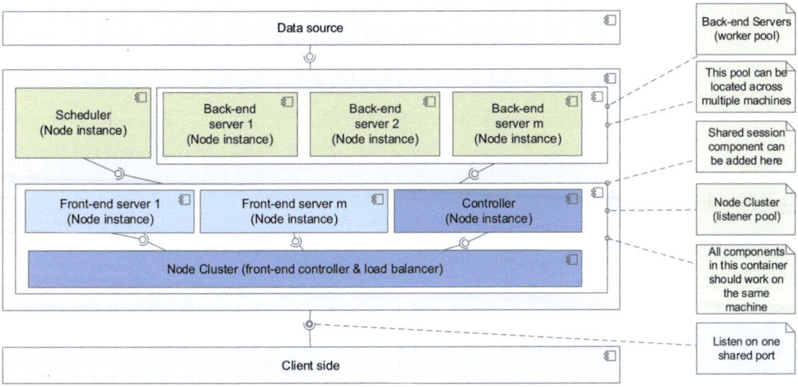

Figure 4: Diagram of the internal structure of Node Scala.

- The **front-end servers** provide the RESTful HTTP(S) interface to the system, handle incoming requests, check their complexity, divide them into sub-tasks, submit the latter for processing, assemble results, and return them to the user. Additional front-end servers can be added to the system as needed, however with the current version of Node Scala using Node Cluster as the front-end controller and load balancer they all have to run on a single machine.

- The **back-end servers** communicate with the database, execute sub-tasks and transfer the results back to front-end servers. Again, additional back-end servers can be added as needed – this time possibly on multiple machines.

- Finally, the **scheduler** keeps track of available back-end servers and assigns them to tasks as requested by the front-end layer.

With the exception of the aforementioned command channel, all the components of Node Scala communicate by streaming JSON data over TCP connections.

4 KAGLVis

Our client application is a data browser which displays selected observables as 3D points at correct coordinates on a virtual globe. At present it can display orbital paths of Envisat as well as cloud altitude measured by MIPAS, in the latter case allowing the user to specify criteria defining clouds. The data is fetched from the server in the background, depending on the user's preferences either set by set or simultaneously for the whole selected range. The view, which can be either a sphere, a plane or that of poles and which allows for selection of a number of different Earth images as background, can be freely rotated and zoomed. The color map in the legend is drawn dynamically on a HTML5 canvas element and synchronized with the contents of the 3D view. Finally, widgets of the user interface use the popular Bootstrap library.

The internal logic of `KAGLVis` has been implemented using AngularJS, with each component of the view assigned its own controller. Dedicated internal services have been created for the exchange of messages between components (no communication through other channels such as global variables is allowed), accessing configuration, and interfacing with the REST server.

The heart of `KAGLVis`, the 3D display (see Figure 5), is based on WebGL Globe – a lightweight JavaScript virtual globe created by Google Data Arts Laboratory which can display longitude-latitude data as spikes. As the name suggests, Globe uses WebGL to leverage local GPU power for 3D rendering. Our version of Globe has been customized to support other views than the original sphere as well as selection of the background texture, reduce memory consumption at a cost of disabling certain visual effects, and most importantly to allow caching of previously displayed data set.

As a single-page application, our client adds only minimal load to the server hosting it. We have therefore deployed it on the same HTTP server as the REST API, reducing overall complication of the architecture.

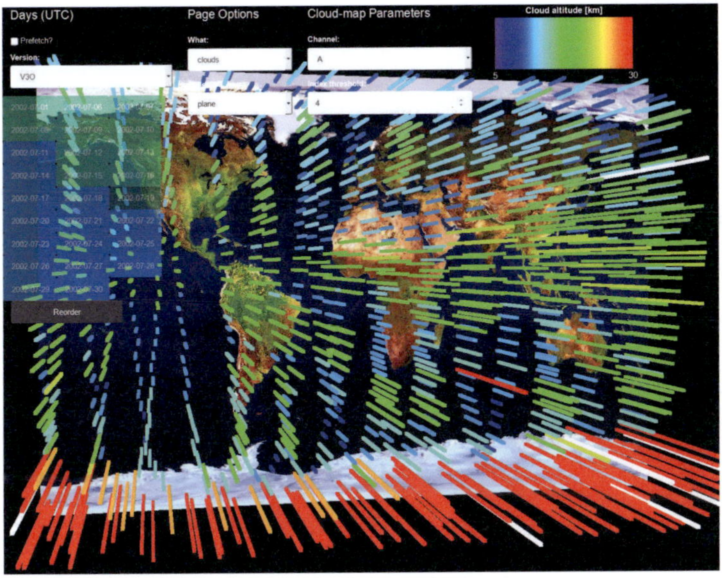

Figure 5: A screenshot of `KAGLVis` showing cloud altitude measured by MIPAS against flat Earth view.

Our evaluation has shown `KAGLVis` to perform well even when subject to considerably more load than encountered during typical use, as well as on machines with modest processing power and only integrated graphics chipsets [Szu16].

5 Conclusion and Contributions

Our Web application-based system for the analysis of environmental satellite data follows the new paradigm of on-the-fly parallel pre-processing of semi-structured data from distributed databases, by RESTful Web services. It is highly distributed, offers multiple degrees of scalability and simplifies deployment of the client. By having based our system on the MEAN stack we have not only been able to take advantage of state-of-the-art, performance-oriented components, which has been reflected by excellent results for performance benchmarks, but also considerably reduced the development overhead. In the course of our work on this system we have:

- migrated metadata of Envisat MIPAS from `MySQL` to `MongoDB` and created `MongoDB` databases for other devices. Among other benefits this migration empowered the usage of parallel access to the datasets of the applications, resulting in an order-of-magnitude performance boost in applications such as geomatching between experiments;
- developed an unique platform allowing parallel processing in Node.js applications;
- designed, deployed and benchmarked a complete MEAN stack-based system for accessing MIPAS data through a Web browser.

Quoted LSDMA Publications

[Ame14] Parinaz Ameri et al. "On the Application and Performance of MongoDB for Climate Satellite Data". In: *13th IEEE International Conference on Trust, Security and Privacy in Computing and Communications, TrustCom 2014, Beijing, China, September 24-26, 2014*. 2014, pp. 652–659. DOI: 10.1109/TrustCom.2014.84. URL: http://dx.doi.org/10.1109/TrustCom.2014.84.

[Maa15] Ahmad Maatouki et al. "A Horizontally-Scalable Multiprocessing Platform Based on Node.js". In: *2015 IEEE TrustCom/BigDataSE/ISPA, Helsinki, Finland, August 20-22, 2015, Volume 3*. 2015, pp. 100–107. DOI: 10.1109/Trustcom.2015.618. eprint: arXiv:1507.02798[cs.DC]. URL: http://dx.doi.org/10.1109/Trustcom.2015.618.

[Szu16] M. Szuba et al. "A Distributed System for Storing and Processing Data from Earth-Observing Satellites: System Design and Performance Evaluation of the Visualisation Tool". In: *2016 16th IEEE/ACM International Symposium on Cluster, Cloud and Grid Computing (CCGrid)*. May 2016, pp. 169–174. DOI: 10.1109/CCGrid.2016.19. eprint: arXiv:1511.07693[cs.DC].

Web Processing Services for the Climate Science Community supported by Birdhouse

Carsten Ehbrecht[a], Stephan Kindermann[a], Jörg Meyer[b]

[a] German Climate Computing Center (DKRZ), Hamburg
[b] Karlsruhe Institute of Technology (KIT), Karlsruhe

Abstract Climate analyses often require the processing of large data from climate model simulations and observation data. Nowadays, researchers download data from climate data archives and try to process those files at their home institutes with local analysis software. But with the growing amount of climate data this is not a feasible solution. With the upcoming climate model simulations projects (e.g. CMIP6) even larger climate institutes and computing centers will not be able to keep all relevant data on one storage system.

Climate processing services can be a valuable contribution to cope with the growing data challenge. A set of climate analyses processing tools can be installed close to the climate data archives. These processing tools are provided as services, which can be accessed via the web. Here, standardized interfaces are important to enable processing services among several institutes interested in climate data. Furthermore a standard processing interface enables the chaining of processes cross-institutional.

Besides sharing the processing services, this gives also the opportunity to share the knowledge (and software) on climate processing among researchers and institutes. By this "reinvented wheels" can be reduced and the software quality can be enhanced. This does not restrict the researchers from choosing their favorite processing tools. Using a standard processing service interface decouples the processing tools (including programming languages, operating systems) from the service itself, and there are also several implementations of the service provider software available.

In Birdhouse we show how the Web Processing Service (WPS) standard can be used to realize services for climate processing tools. WPS is an open standard defined by the Open Geospatial Consortium (OGC) with several open source implementations. In Birdhouse we currently use the Python implementation of WPS, "PyWPS", but Birdhouse is not restricted to a single WPS implementation. Birdhouse is **not** yet another processing framework, Birdhouse provides "glue" and missing parts to successfully run WPS for climate data processing.

1 Introduction

A Web Processing Service (WPS) [15e] is a standard defined by the Open Geospatial Consortium (OGC) to provide processing capabilities as a web-service. The standard defines operations to discover and execute processes on a web-service by a client, and how the inputs and outputs of a process are specified (see Figure 6). The WPS operations are:

- `GetCapabilities` - list all available processes with identifier, title and abstract. This operation is used to discover the capabilities of a WPS and to bind them to a WPS client.
- `DescribeProcess` - show the details of a specific process with input and output parameters (including data types).
- `Execute` - run a process with user defined input parameters. Short running processes (few seconds) can be run synchronously and long running processes asynchronously (by polling the process status). Process outputs can be returned directly or stored on server side and returned by a URL reference.

WPS defines a simple HTTP interface, which can be accessed with `GET` and `POST` requests. WPS requests can be initiated directly by URLs, scripts and

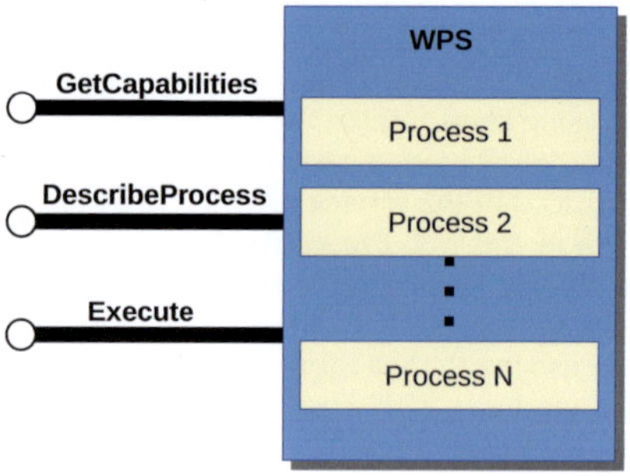

Figure 6: Web Processing Service Operations.

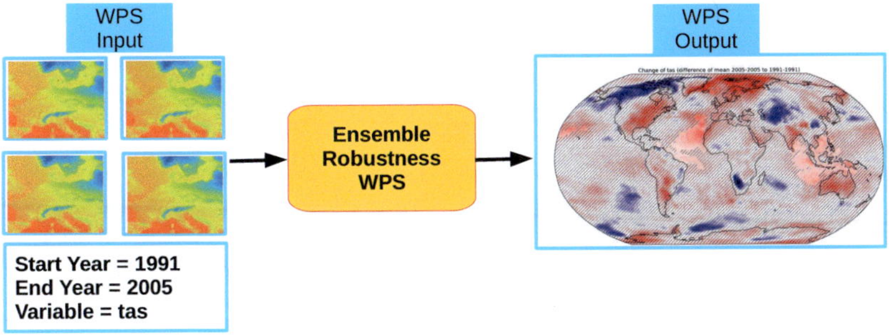

Figure 7: WPS process example for a statistical robustness calculation of ensemble input data.

WPS aware clients (web-portals, GIS desktops). WPS is, like any other OGC service, a state-less protocol. There is no communication history between the server and the client. WPS processes can be chained, either manually or by a workflow-engine. Processing data can be either local on the server side or supplied by input parameters (`WPS ComplexType`).

Birdhouse [16d] is a collection of Web Processing Service related components to support data processing in the climate science community. Birdhouse supports setting up your own WPS services and provides generic WPS clients to interact with them. In addition it comes with the integration of interfaces to climate data archives like Earth System Grid Federations (ESGF) [11c] and Thredds data catalogs [16w].

Figure 7 shows an example how a WPS process is executed, with an ensemble of climate data in NetCDF format (surface temperature) as input and an image with a statistic of the ensemble robustness as output.

2 Birdhouse Components and Architecture

The Birdhouse ecosystem includes components to access and catalog external data sources provided by Thredds data servers including the Earth System Grid Federation (ESGF) data archive. Additionally it includes a component to index and search large local data collections using Solr [04] technology.

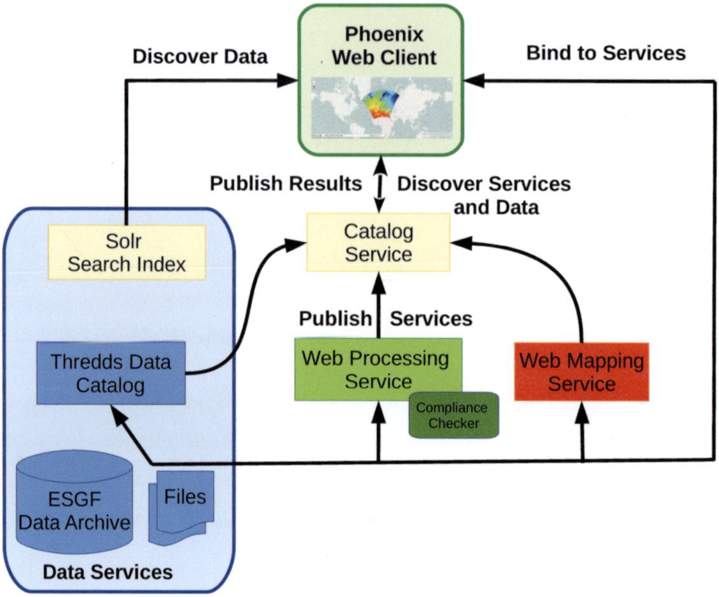

Figure 8: Birdhouse infrastructure with WPS, WMS, Catalog Service and climate data archives.

For managing and interacting with processing services Birdhouse uniformly exposes OGC WPS standard based interfaces. The OGC WPS interface descriptions can be registered in an OGC Web Catalog Service [13c] supporting standards based service discovery. Processing results can be published to the same Catalog Service (see Figure 8).

Birdhouse has a web-client called "Phoenix" to interact with Web Processing Services and to feed them with data from climate data archives.

To setup the WPS infrastructure Birdhouse uses a build system based on Conda [16h], Buildout [06b] and Ansible [14a].

To control the user access to WPS services (and other OGC services) Birdhouse has a OWS security proxy "Twitcher", which can be placed in front of any WPS service.

3 Climate Processing Services

Birdhouse has several Web Processing Services, which combine processes of different aspects in climate data processing. Currently these are:

- **Flyingpigeon** contains a variety of processes ranging from simple polygon subsetting to complex data analysis methods and workflows used in climate impact or extreme weather event studies [16x].
- **Hummingbird** provides processes to check conformance to climate metadata standards. Theses standards are the NetCDF-CF (Climate and Forecast conventions) and metadata conventions of climate data simulation projects like CORDEX and CMIP6.
- **Malleefowl** has processes to access climate data archives like Earth System Grid Federation (ESGF) and Thredds data catalogs. It includes a workflow process to fetch climate data from a selected archive and provides this data to a selected analysis process. If the requested climate data files are not already locally available on disk then they will be downloaded and cached on file-system.
- **Emu** has some lightweight processes to show which input and output parameters are supported by WPS and to provide examples to write your own processes.

4 Birdhouse Build System

Birdhouse consists of several components like Flyingpigeon, Emu and Phoenix. Each of them can be installed individually. The installation is done using the Python-based build system "Buildout". Most of the dependencies are maintained in the Python distribution system "conda". For convenience each Birdhouse component has a Makefile to ease the installation so one does not need to know how to call the Buildout tool.

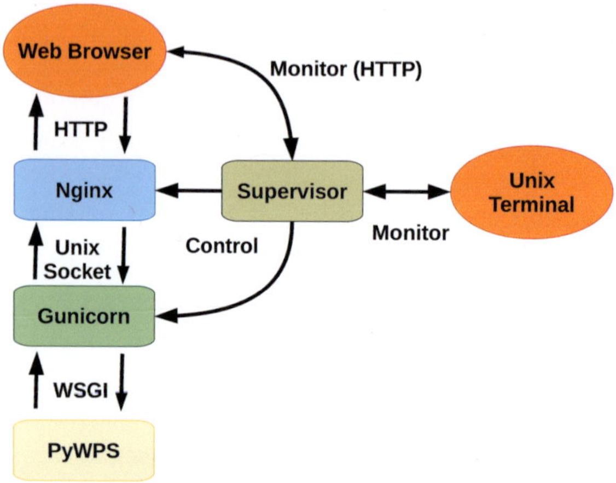

Figure 9: PyWPS [08] WSGI Application.

We use the Gunicorn HTTP application server (similar to Tomcat for Java servlet applications) to run the Birdhouse web applications with the WSGI interface. In front of the Gunicorn application server we use the Nginx HTTP server (similar to Apache web server). All these web services are started/stopped and monitored by a Supervisor service (see Figure 9).

5 Phoenix

Pyramid Phoenix is a web-application build with the Python web-framework Pyramid [11g]. Phoenix has a user interface to interact with Web Processing Services. The user interface allows you to register Web Processing Services. For these registered WPS services you can list the available processes. You are provided with a form page to enter the parameters to execute a process and you can monitor the jobs and see their results (see Figure 10).
In the climate science community many analyses use climate data in the NetCDF format. Phoenix uses the Malleefowl WPS which provides processes to access NetCDF files from the ESGF data archive. Malleefowl provides a workflow process to chain ESGF data retrieval with another WPS process,

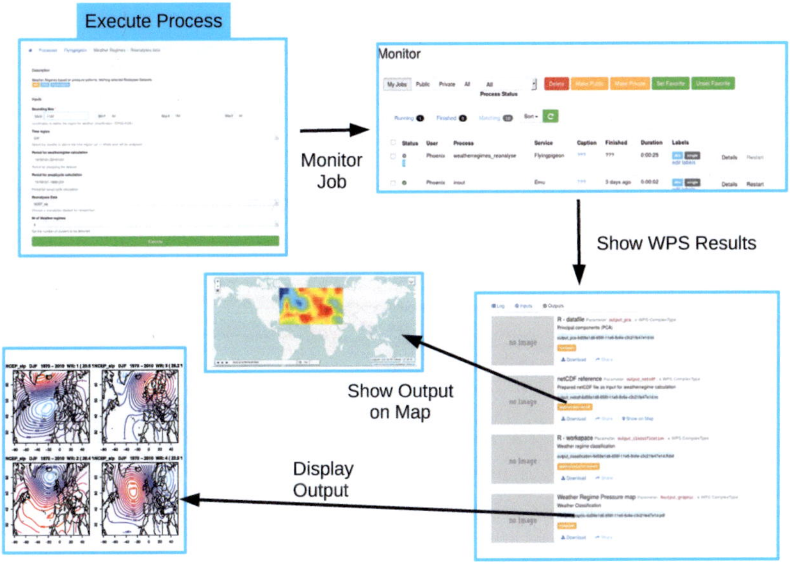

Figure 10: Phoenix example with the execution, monitoring and displaying outputs of a WPS processing job.

which needs NetCDF data as input. Phoenix has a Wizard to collect the parameters to run such a workflow with a process of a registered Web Processing Services.

Phoenix is intended to help researchers and developers of WPS processes to use their processes more conveniently, especially for feeding their processes with different data sources (like ESGF data archive).

Phoenix visualizes processing input and output data in the NetCDF format on a map (based on Leaflet). The climate data map is generated by a Web Mapping Service (WMS). WMS supports a time attribute, which can be used to step through the map by time.

In addition one can use the Birdy command-line tool to work with Web Processing Services.

6 Twitcher

Twitcher is a security proxy for Web Processing Services. The execution of a WPS process is protected by the proxy. The proxy service provides access tokens (uuid strings), which need to be used to run a WPS process. The access tokens are valid only for a short period of time (see Figure 11).
The implementation is not restricted to WPS services. It will be extended to more OWS services like WMS (Web Map Service) and CSW (Catalogue Service for the Web) and might also be used for Thredds catalog services.
Twitcher consists of the following parts:

- **OWS Security** is a WSGI middleware, which puts a simple token based security layer on top of a WSGI application. The access tokens are stored in a MongoDB.
- **OWS Proxy** is a WSGI application, which acts as a proxy for registered OWS services. Currently it supports WPS and WMS services.
- **Administration interface** is a XML-RPC service which is used to control the token generation and OWS service registration. The interface is accessed using "Basic Authentication". It is meant to be used by an administrator and administrative web portals.

The OWS security middleware protects OWS services with a simple string based token mechanism. A WPS client needs to provide a string token to access the internal WPS or a registered OWS service. A token is generated by using the XML-RPC admin interface. This interface is supposed to be used by an external administration client which has user authentication and generates an access token on behalf of the user.
The OWS security middleware allows by default to use `GetCapabilities` and `DescribeProcess` requests without a token. The `Execute` request can be accessed only with a valid token.

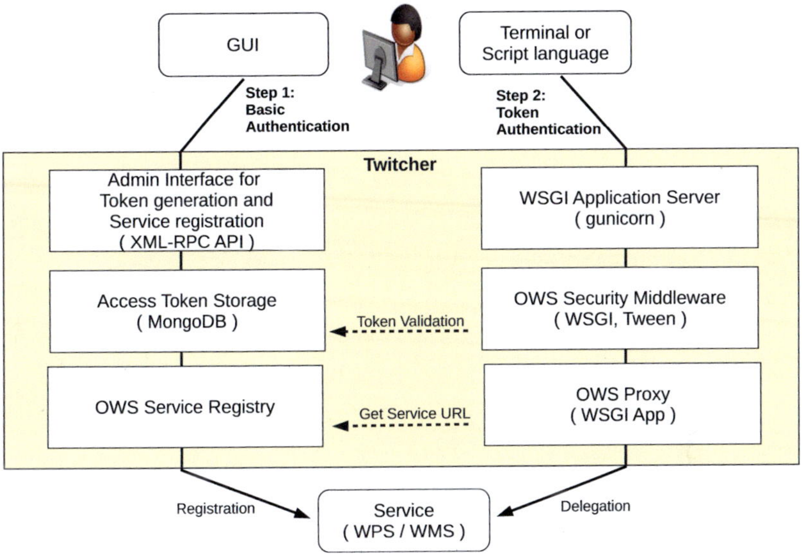

Figure 11: Twitcher security proxy for OWS services.

Twitcher is meant to be integrated in existing processing infrastructures with OGC/OWS services and portals and can be put in front of any WPS service.

7 Conclusions and Contributions

Further work has to be done on the security infrastructure of WPS servers. The WPS standard does not provide a solution besides using HTTPS and securing the WPS server with username/password. Currently we are using the Twitcher security proxy with simple access tokens passed as HTTP header parameters. In a distributed installation we need tokens which contain security information that can be verified on a remote Twitcher service, Macaroons [14b] might be a promising option.

In the current deployment of a WPS service processes are directly executed on the machine where the WPS service is running. In installations on computing centers we need to attach existing batch job processing systems like Slurm to make use of the existing compute facilities. Another task is to use docker containers for the execution of WPS processes to encapsulate each processing job.

In addition to this the next steps include the collaboration with European partners to make WPS services interoperable and usable in international collaborations.

This use case demonstrates the necessity for a close collaboration between researchers and data scientists coming from different institutions. Also it shows the large variety of services and tools involved in this collaboration.

Quoted LSDMA Publications

[16d] *Birdhouse Open Source Project, Collection of WPS related components to support Climate data processing*. 2016. URL: http://bird-house.github.io/.

Other References

[04] *Solr, Open Source search engine*. 2004. URL: https://lucene.apache.org/solr/.

[06b] *Buildout, Python Software Build System*. 2006. URL: http://www.buildout.org/en/latest/.

[08] *PyWPS, Python implementation of the Web Processing Service standard from the Open Geospatial Consortium*. 2008. URL: http://pywps.org/.

[11c] *Earth System Grid Federation*. 2011. URL: http://esgf.llnl.gov/.

[11g] *Pyramid, Python web framework*. 2011. URL: http://www.pylonsproject.org/.

[13c] *PyCSW, Python implementation of an OGC catalog service for the web*. 2013. URL: http://pycsw.org/.

[14a] *Ansible, IT automation tool in Python*. 2014. URL: https://www.ansible.com/.

[14b] *Macaroons, authorization credentials that support decentralized verification*. 2014. URL: https://github.com/rescrv/libmacaroons.

[15e] *The OpenGIS Web Processing Service (WPS) Interface Standard*. 2015. URL: http://www.opengeospatial.org/standards/wps.

[16h] *Conda, Open Source package management system*. 2016. URL: http://conda.pydata.org/docs/.

[16w] *Thredds Data Server*. 2016. URL: http://www.unidata.ucar.edu/software/thredds/current/tds/TDS.html.

[16x] *Web processing service for climate impact and extreme weather event analyses, Nils Hempelmann*. 2016. URL: https://hal.archives-ouvertes.fr/hal-01375615v2.

A Concept for a User-oriented Energy Data Management System

Fabian Rigoll[a], **Hartmut Schmeck**[a]

[a] Karlsruhe Institute of Technology (KIT), Karlsruhe

Abstract The amount of energy-related data is increasing: Smart meters, smart plugs, and even household appliances themselves create detailed records of energy data. This kind of data can be used to gain valuable insights into the status of the energy grid. Furthermore, energy data can serve as a basis for elaborate optimization and for additional services, leading to a significant commercial value. But at the same time, energy data often is highly sensitive data that can affect residents' privacy. Therefore, an energy data management system is needed which satisfies both technical as well as user-driven requirements. This paper presents a concept for such a user-oriented energy data management system. During the design phase, a detailed requirements analysis was performed. An initial data life cycle analysis focussed on technical aspects. Ensuing this step, an analysis from a user's perspective was performed. The resulting requirements form a list of specifications which were used to create a modular concept called *Data Custodian Service*. This system has also been implemented in form of a demonstrator.

1 The Energy Transition and Energy Informatics

Humanity depends on many different kinds of energy. People in a modern society are accustomed to be able to use gas, heat, and electricity at almost any given time. A reliable energy supply is taken for granted and we are able to make use of energy without much further consideration. However, if a power failure – a so-called blackout – occurs, we are reminded of how much we depend on energy. Due to this subliminal consumption, we don't realize the vast amounts of energy which are necessary for our day to day life. If for example a washing machine were to be fueled not by electricity but by the combined energy output of humans, the constant power of twenty people would be necessary for a normal washing cycle: An averagely trained person can provide some one hundred watts for a longer period of time and a washing machine has a peak demand of roughly two thousand watts. And yet, our electricity grid reliably provides energy for millions of households and industry on demand. This thirst for energy is ever-increasing. Devices and appliances are becoming more efficient, but at the same time Earth's population is rapidly growing: the annual increase in population is about 80 million people [Sta16]. In the course of their lives, all of these people will use great amounts of energy which has to be provided somehow.

In the past, fossil fuels such as oil, coal, and gas were virtually the only energy source. However, using such resources implies severe repercussions. By pumping up oil and by mining coal, entire regions are affected or even destroyed. Furthermore, the human-caused greenhouse effect has irreversibly increased Earth's temperature [Coo13]. The concentration of carbon dioxide in Earth's atmosphere – one of the most important drivers of the greenhouse effect – is at the maximum of the last 650 000 years [Qua15]. Global warming cannot be stopped anymore, it can only be slowed down and in order to limit the temperature rise, a massive reduction in the emission of greenhouse gases is necessary. This means that the use of fossil fuels must be reduced significantly while at the same time expanding the use of renewable energy resources such as solar and wind power.

The process of abolishing fossil fuels in favor of renewable energies is often called *Energiewende* or *energy transition* and is considered one of this century's greatest challenges [AB07]. In a classical energy grid there is a small number of central mostly fossil-fueled power plants which provide energy for the consumers. If demand increases, more energy is fed into the grid. However, power plants which utilize renewable energies usually can't increase their energy output to match the demand – a wind turbine can't provide elec-

trical power if there is no wind. This fluctuation has to be dealt with. Thus, supply won't follow demand anymore, but demand will have to follow supply – at least to some degree. Additionally, instead of a few hundred central power plants many millions of decentral entities will have to be controlled. Thus, a profound change in the entire energy system is due. Efforts in many different areas of research are needed for a rapid and successful transformation. One of these is the area of Energy Informatics. Its main goal is the efficient integration and use of renewable energy resources and the most effective utilization of demand flexibility by employing information and communication technology (ICT) in the energy system [Goe14]. Using ICT offers the possibility to make the grid and its operation smarter by gaining a deeper insight into the energy system. In this manner, the energy system can be made more reliable and efficient than would be possible by mere engineering.
Information gained in the energy system in form of energy data is the basis for deliberate decisions and optimization. The recorded energy data can be used to improve various aspects of the grid itself and the processes therein. However, energy data often is highly sensitive data that can potentially compromise people's privacy or a company's secrets. Therefore, a suitable energy data management system is necessary that protects the privacy of its users while at the same time preserving the usability of the data for the smart grid.

2 Energy Data

In this section, the term energy data is defined. Afterwards, benefits which arise from using energy data as well as potential privacy threats are discussed.

2.1 Definition

In the literature, there are different notions of the term energy data. Often, data from smart meters are referred to as energy data (e.g. [Mol10]). In other cases, it is not only data from smart meters but also data from the appliances

themselves (e.g. [KJ11]). The previous examples refer to electrical energy. However, there are also meters for gas consumption [FK10]. This small excerpt shows how ambiguous this term can be. Not only the data sources differ, but also data formats, ways of transmission etc.
Evidently, the term energy data cannot be restricted to one specific application. Rather, a broad understanding of the term is necessary. Therefore, in the following, all data are considered to be energy data as long as the following two conditions are met:

- The data comprise information with reference to the use (production, consumption, conversion, ...) of energy.
- The data are available in form of a time series.

This definition is deliberately wide so that many different kinds of energy data can be subsumed. At the same time, other related data are excluded: energy model data are not a part of this definition as they are not time series data.

2.2 Benefits of Energy Data

Energy data can be used to gain detailed insight into the energy grid and its status. This makes energy data valuable and often highly sensitive. In the following, a few examples of possible applications where energy data can help to implement and speed up the energy transition are given.
For many people, using energy is a rather abstract process. If no direct feedback is available, the amount of energy used often cannot even be guessed. Therefore, consumers usually don't know how an adjustment in behavior would affect their energy bill. As soon as consumers do get feedback however, up to 15 % of reduction in consumption are possible – without any further optimization [Dar06]. If an energy management system is used to optimize the operation of appliances in a household, even higher savings can be expected. It is conceivable that further novel services will arise, based on households' energy data, e.g. consumption assessment or advice on how to

reduce overall cost. In such instances, users might profit by sharing their data with such service providers or other third parties who are interested in the data. At the same time, those providers profit by gaining access to valuable data.

As explained before, in the future energy, consumption will have to follow supply to some extent. To achieve this, energy management systems are becoming necessary. In order to be able to make reasonable decisions, they need energy data. The availability and quality of such data significantly influences the quality of the optimization in the smart grid [Fan13]. Detailed energy data can also help to better understand consumption patterns and therefore improve the quality of forecasts which in turn can support optimization efforts [KD12].

Those are just some of the use cases where energy data can be used to improve the smart grid and its efficient operation. However, energy data also pose a severe threat to people's privacy.

2.3 Privacy Threats

More than two decades ago Hart [Har92] published his pioneering paper on nonintrusive load monitoring. The underlying idea is that not only electricity is transported when consuming electrical energy, but also information: If the measured consumption patterns are interpreted correctly, detailed information from within a household can be revealed.

Many improved and novel approaches have been published in the past years [ZR11]. Even though the technical side might not be of interest to the average consumer, the consequences affect them nonetheless:

- How many people are living in a household?
- A person called in sick at work. Did that person stay at home on that particular day?
- Do the residents in a home sleep well or do they get up at night?

- How and when do the residents prepare their meals?
- Did a person leave in time for work?

Neither very detailed consumption data on a device level nor high resolutions are necessary to answer these questions. Energy data at an aggregated household level with a temporal resolution in the area minutes is sufficient [Mol10]. If more fine-grained data is available, an analysis can even reveal which TV content is consumed at a given time in a home [GJL12]. Many other aspects such as behavioural patterns, religion etc. can be detected with surprisingly high precision.
If an adversary gained access to energy data of a household, the residents' privacy could obviously be compromised. Therefore, energy data must be well protected. However, in some cases residents might still wish to share some data. Therefore, a suitable energy data management system is necessary which helps the users to decide whether to share data and at which quality. Similarly, enterprises will be reluctant to provide their energy data because they are afraid to reveal essential information about various kinds of internal processes.

3 Requirements Analysis

In order to account for technical as well as user-driven aspects, a detailed multi-step analysis of energy data's properties was performed. The resulting requirements are the basis for the concept presented in this paper.

3.1 Technical Data Life Cycle Analysis

In a first step, the data life cycle of energy data was analyzed. A typical data life cycle consists of the following six phases: acquisition, transmission, storage, analysis, distribution, and deletion. For each of these steps, the specifics of energy data were considered. This systematic approach ensures that the entire life cycle is covered. Resulting requirements are among others:

- Support for different kinds of energy data, regardless of file formats
- Robustness in relation to faulty or missing data
- Easily searchable meta data
- Efficient storage for large amounts of time series
- Efficient and flexible access at different resolutions

Different sources of energy data lead to different kinds of energy data. However, all relevant data is usually available in some form of time series. Therefore, the energy data management system should not restrict itself to one data format but it should be able to import different kinds of time series based data. Many energy metering systems use wireless connections. Therefore, transmission errors in form of faulty or missing data can occur. The energy data management system should be able to deal with this situation. As with most data, energy data needs to be efficiently searchable. The users must be able to filter by device or time, for example. Even if energy data in small households might not appear to be large data, it is recorded constantly over possibly long periods of time. Therefore, an efficient way of storing it is necessary. In order to efficiently use the energy data, flexible access at various different resolutions is needed. The data life cycle analysis revealed some other requirements which are skipped here for brevity, but have been included in the published PhD thesis [Rig17].

3.2 User-Oriented Analysis

In addition to the technical analyses, the privacy implications of energy data were investigated. To that end, an analysis of nonintrusive load monitoring techniques was conducted: What information can actually be derived from energy data and how is that done?

There is a large number of different approaches for various situations. Some of them work only for fine-grained data but deliver excellent results in terms of device recognition [GRP10]. Others don't need high resolutions and often can't identify individual devices but reliably reveal presence or absence of residents [Mol10]. Temporal as well as spatial resolutions play a crucial role in the quality of the results, but even very coarse data can lead to an invasion of residents' privacy. That is why smart meters might even be "regarded as a judashole into a household" by some [KS14]. Following the analysis, these requirements were defined among others:

- No distribution of energy data or derived data without explicit consent by the user(s)
- Reduction of data quality to the least acceptable degree before any data is shared if possible
- Explanation of the privacy implications if data were to be shared

From a user's point of view the distribution of energy data must be restricted rigorously in order not to jeopardize their privacy. If the users decide to share data all the same, at least the data quality should be lowered as much as possible. Data with a low resolution and artificial noise might reveal less than high quality data. The privacy aspect reveals another issue: the users' lack of knowledge. If a user is not able to understand the data itself (cf. [Dar06]), it is highly unlikely that the privacy implications are clear. Therefore, an explanation of the privacy implications is necessary.

At a time where credit cards and smartphones are used on a daily basis, one might argue that energy data is a lesser evil in comparison. However, credit cards and smartphones are opt-in. It might come with a loss of comfort, but it is possible to live without them [Sta10]. The use of (electrical) energy is not really optional. It might be possible in very extreme cases but comes with a drastic loss of comfort. As the users often can hardly decide whether energy data is collected, they should at least be given the chance to decide what is done with their data. Therefore, the user-driven requirements may not be omitted when designing an energy data management system.

3.3 Summary

In their entirety, the above requirements form a specification which can be used to design an energy management system that will not only comply with technical demands but will also satisfy user needs. Both aspects are equally important. The energy data management system must work efficiently with all kinds of energy data so that the need does not arise to use other applications over which the users don't have control. At the same time, the energy data should be put to good use while at the same time protecting the users' privacy as effectively as possible.

4 Concept

In this section, the architecture of the *Data Custodian Service* (DCS) is described (cf. Figure 12). Since its first publication it has been modified and improved [Rig14; RS14]. It is a modular system which has been designed to comply with the requirements of the carried out analysis. The DCS consists of several different modules which interact in order to support the management of energy data in all six phases of the data life cycle. In the following, the individual parts of the system are characterized by describing typical tasks.

4.1 Data Ingest

Data sources can send energy data to the DCS by using a *Machine-to-Machine Interface* – a webservice – offered by the *DCS webserver*. The *Data Custodian Core* – the central module of the DCS – then hands over the transmitted data to the *Time Series Handler* which processes the data and converts the different kinds of data formats to a consistent internal format. As shown earlier, there is a large number of diverse *Data Sources* that need to be incorporated. By using a dedicated *Time Series Handler* which is capable of converting a variety of different input formats, all kinds of energy data sources can be supported.

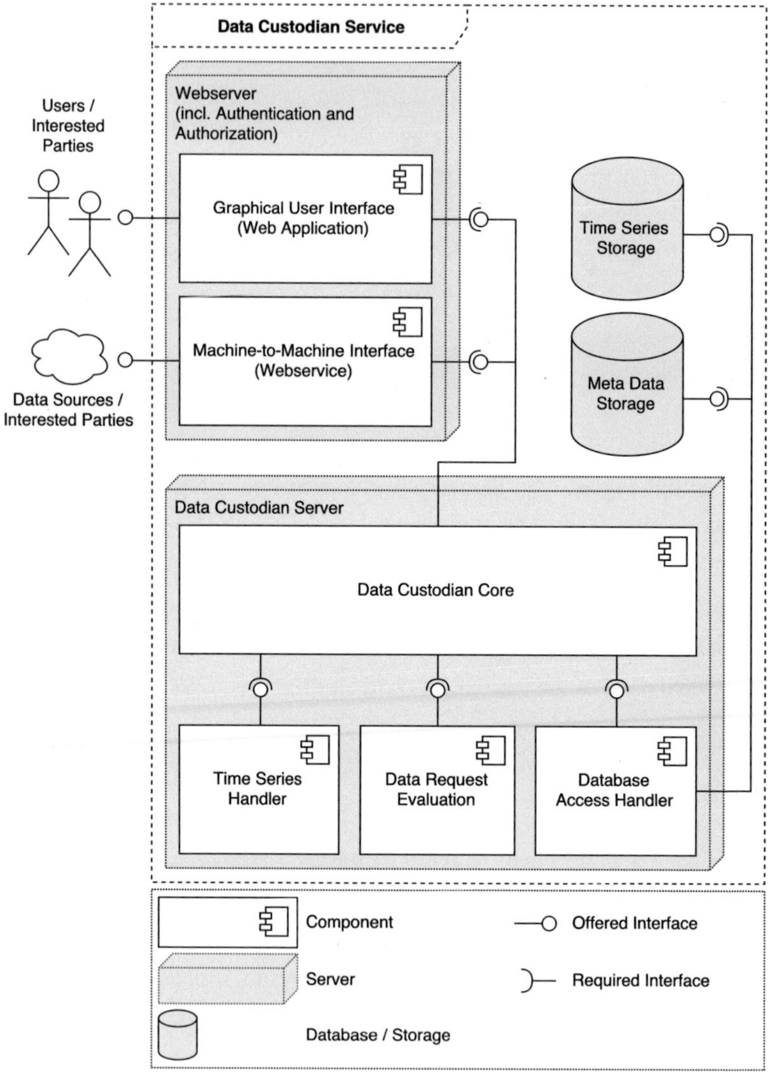

Figure 12: The architecture of the Data Custodian Service (DCS).

4.2 Data Storage

The unified data are handed back to the *Data Custodian Core* after they have been preprocessed in the *Time Series Handler*. The *Data Custodian Core* uses the *Database Access Handler* to create a new entry in the *Meta Data Storage* and to put the energy data itself into the *Time Series Storage*. The *Meta Data Storage* needs to be easily searchable, so that stored data can be filtered efficiently by criteria such as data source or time span. The *Time Series Storage* must be – as its name implies – an efficient storage for large amounts of time series data. However, the actual energy data set identification is done via the corresponding entry in the *Meta Data Storage* from which it can be referenced and then retrieved from the *Time Series Storage*.

4.3 Data Requests

For obvious reasons *Interested Parties* cannot be allowed to access any data directly. Instead, they have to issue a request for data. The *Graphical User Interface* is a web application which offers an interface for *Interested Parties* to issue such requests. Incoming data request are transferred to the *Data Custodian Core* which uses the *Data Request Evaluation* module to assess the potential privacy implications of the request. Among others, both the temporal and spatial resolutions as well as the requested period of time influence the assessment. The evaluation is done by a module so that it can be swapped out easily in order to allow for different kinds of evaluation schemes. A summary of the data request and the result of the request evaluation is presented to the *User*. The *User* – as the one who is affected by the energy data – must then decide, whether that data request should be accepted or declined. The privacy evaluation ensures that the *User* knows about potential consequences when sharing the data. In case the *User* declines the data request, the *Interested Party* is informed and no data is exported. If the *User* accepts the request, the *Data Custodian* fetches the times series data from the corresponding *Time Series Storage* using the reference in the *Meta Data Storage*. The data are then processed according to the data request (selection, quality reduction etc.). In a last step, the derived data is sent to the *Interested Party*. By employing such a system, the *User* is always informed about any outgoing data and the potential consequences for the residents' privacy.

4.4 Logging

The *Data Custodian Core* automatically logs all events relevant to the users' privacy in the *Meta Data Storage*. That even includes incoming data requests by *Interested Parties* which are ignored by the user or declined. This allows for the recognition of potential abuse patterns which might occur. The tracking of approved data requests ensures that potential data abuse can be mapped to a *Third Party* who requested the corresponding data. This facilitates prosecution in case of data abuse.

5 Demonstrator

The presented concept in form of the *Data Custodian Service* has been implemented in form of a demonstrator using *Python*. This section gives an overview of the used software.

The *Python* web framework *Django*[1] is at the core of the demonstrator. *Django* works as a model-view-presenter: It serves as a user interface and manages the underlying energy data sets' meta data which are stored in an *SQLite*[2] database. The demonstrator offers a web service and web application through which new data sources can be added. Furthermore, the web interface allows to issue and manage incoming data requests. Consequently, it is the contact point for both interested parties and users (the owners of the energy data).

[1] https://www.djangoproject.com/
[2] https://www.sqlite.org/

While the meta data are stored in an *SQLite* database by *Django*, the actual time series data are stored in *HDF5*[3] files. This file format specializes in the efficient storage of homogeneous tables and as time series can be represented as such, it is well suited for storing them. At the same time, efficient selection of time ranges is possible, thus offering flexible access to different sections of the data. Referencing stored energy data is done by *Django* using the corresponding meta data.
The time series data are handled using the data analysis library *pandas*[4]. This library is capable of handling large amounts of time series data in so-called *DataFrame* objects allowing for efficient selection, resampling, and other manipulation of data. *pandas* is used to provide the functionality of the *Time Series Handler* as well as the analysis of the time series data.

6 Conclusion and Contributions

During the work within the *Data Life Cycle Lab Energy* the need for a user-oriented energy data management system arose. In order to fulfill both technical as well as user-driven requirements, an extensive analysis was performed. On one hand, a technical data life cycle analysis was performed. This step yielded several requests that must be met from a technical point of view. On the other hand, a user-oriented analysis was carried out, including a detailed analysis of potential privacy threats arising from energy data. This led to a number of requirements which must be satisfied by a suitable energy data management system from a user's perspective.
Progressing from this set of requirements, a concept for an energy data management system was designed. During the design phase, both technical as well as user-driven demands were addressed. The work which was briefly presented in this paper has been published in form of a PhD thesis [Rig17].

[3] https://www.hdfgroup.org/HDF5/
[4] http://pandas.pydata.org/

Quoted LSDMA Publications

[Rig14] F. Rigoll et al. “A Privacy-Aware Architecture for Energy Management Systems in Smart Grids”. In: *Ubiquitous Intelligence and Computing, 2014 IEEE 11th Intl Conf on and IEEE 11th Intl Conf on and Autonomic and Trusted Computing, and IEEE 14th Intl Conf on Scalable Computing and Communications and Its Associated Workshops (UTC-ATC-ScalCom).* Dec. 2014, pp. 449–455. DOI: 10.1109/UIC-ATC-ScalCom.2014.9.

[Rig17] Fabian Rigoll. “Nutzerorientiertes Energiedatenmanagement”. PhD thesis. Karlsruhe Institute of Technology, 2017. DOI: 10.5445/IR/1000068109.

[RS14] Fabian Rigoll and Hartmut Schmeck. “Konzeption eines Energiedatenmanagementsystems unter Beachtung von Datenschutz und Privatsphäre”. In: *VDE-Kongress 2014*. Ed. by VDE. VDE. VDE VERLAG GmbH, Oct. 2014.

Other References

[AB07] Nicola Armaroli and Vincenzo Balzani. “The future of energy supply: Challenges and opportunities”. In: *Angewandte Chemie International Edition* 46.1-2 (2007), pp. 52–66. ISSN: 1521-3773. DOI: 10.1002/anie.200602373.

[Coo13] John Cook et al. “Quantifying the consensus on anthropogenic global warming in the scientific literature”. In: *Environmental Research Letters* 8.2 (2013), p. 024024. DOI: 10.1088/1748-9326/8/2/024024.

[Dar06] Sarah Darby. *The effectiveness of feedback on energy consumption*. Tech. rep. Environmental Change Institute, University of Oxford, 2006.

[Fan13] Zhong Fan et al. "Smart grid communications: Overview of research challenges, solutions, and standardization activities". In: *Communications Surveys Tutorials, IEEE* 15.1 (2013), pp. 21–38. ISSN: 1553-877X. DOI: 10.1109/SURV.2011.122211.00021.

[FK10] English. In: *Pervasive Computing*. Ed. by Patrik Floréen and Jens Krüker. Vol. 6030. Lecture Notes in Computer Science. 2010, pp. 154–165. ISBN: 978-3-642-12653-6.

[GJL12] Ulrich Greveler, Benjamin Justus, and Dennis Loehr. "Multimedia content identification through smart meter power usage profiles". In: *Computers, Privacy and Data Protection* 1 (2012), p. 10.

[Goe14] Christoph Goebel et al. "Energy Informatics". In: *Business & Information Systems Engineering* 6.1 (2014), pp. 25–31. DOI: 10.1007/s12599-013-0304-2.

[GRP10] Sidhant Gupta, Matthew S. Reynolds, and Shwetak N. Patel. "ElectriSense: single-point sensing using EMI for electrical event detection and classification in the home". In: *Proceedings of the 12th ACM International Conference on Ubiquitous Computing*. UbiComp '10. Copenhagen, Denmark: ACM, 2010, pp. 139–148. ISBN: 978-1-60558-843-8. DOI: 10.1145/1864349.1864375.

[Har92] George William Hart. "Nonintrusive appliance load monitoring". In: *Proceedings of the IEEE* 80.12 (1992), pp. 1870–1891.

[KD12] G. Kalogridis and S. Dave. "PeHEMS: privacy enabled HEMS and load balancing prototype". In: *Smart Grid Communications (SmartGridComm), 2012 IEEE Third International Conference on*. Nov. 2012, pp. 486–491. DOI: 10.1109/SmartGridComm.2012.6486032.

[KJ11] J Zico Kolter and Matthew J Johnson. "REDD: A public data set for energy disaggregation research". In: *Workshop on Data Mining Applications in Sustainability (SIGKDD), San Diego, CA*. Vol. 25. Citeseer. 2011, pp. 59–62.

[KS14] Markus Karwe and Jens Strüker. "A survey on privacy in residential demand side management applications". English. In: *Smart Grid Security*. Ed. by Jorge Cuellar. Vol. 8448. Lecture Notes in Computer Science. Springer International Publishing, 2014, pp. 154–165. ISBN: 978-3-319-10328-0. DOI: 10.1007/978-3-319-10329-7_10.

[Mol10] Andrés Molina-Markham et al. "Private memoirs of a smart meter". In: *Proceedings of the 2Nd ACM Workshop on Embedded Sensing Systems for Energy-Efficiency in Building*. BuildSys '10. Zurich, Switzerland: ACM, 2010, pp. 61–66. ISBN: 978-1-4503-0458-0. DOI: 10.1145/1878431.1878446.

[Qua15] Volker Quaschning. *Regenerative Energiesysteme: Technologie – Berechnung – Simulation*. 9., aktualisierte und erw. Aufl. Hanser eLibrary. München: Hanser, 2015. ISBN: 978-3-446-44333-4. DOI: 10.3139/9783446443334.

[Sta10] R. Stallman. "Is digital inclusion a good thing? How can we make sure it is?" In: *Communications Magazine, IEEE* 48.2 (Feb. 2010), pp. 112–118. ISSN: 0163-6804. DOI: 10.1109/MCOM.2010.5402673.

[Sta16] Statista – Das Statistik-Portal. *Zuwachs der Weltbevölkerung, angegeben in unterschiedlichen Zeiteinheiten (Stand 2016)*. Online. 2016. URL: http://de.statista.com/statistik/daten/studie/1816/umfrage/zuwachs-der-weltbevoelkerung/.

[ZR11] Michael Zeifman and Kurt Roth. "Nonintrusive appliance load monitoring: Review and outlook". In: *Consumer Electronics, IEEE Transactions on* 57.1 (2011), pp. 76–84.

Managing Large Data Sets in Super-resolution Optical Microscopy

Ajinkya Prabhune[a], Nick Kepper[b], Katharina Aschenbrenner[c], Sebastian Butzek[c], Christian Guthier[c], Matthias Krufczik[d], Margund Bach[d], Eberhard Schmitt[de], Michael Hausmann[d], Jürgen Hesser[c]

[a] Karlsruhe Institute of Technology (KIT), Karlsruhe
[b] DKFZ German Cancer Research Center, Heidelberg
[c] Experimental Radiation Oncology, Mannheim Medical Center — University of Heidelberg
[d] Kirchhoff Institute for Physics, Heidelberg University
[e] Institute for Numerical and Applied Mathematics, University of Göttingen, Göttingen

Abstract Localization microscopy, especially SPDM (Spectral Position Determination Microscopy), can scale optical resolution down almost to the electron microscopy level in the 10 nm range, which is important for biological and medical research and diagnosis. But these techniques produce image data in the range of GB/s and require the handling, processing and evaluation of image stacks of up to thousands of frames per single cell. These data have to be stored and made accessible for the research community, in diagnostic connotation for 30 years by law. To this end, we have designed a system for transmitting the data, adding meta data for characterization and retrieval, storing the data, and also offering programs and processing procedures for fast evaluation, on individual machines or in clusters. A Generic Client Service (GCS) API for connecting disparate services is designed and implemented seamlessly integrating with the KIT Data Manager and the Large Scale Data Storage. A structured metadata model based on Core Scientific Metadata Model (CSMD) is established for describing the extremely large datasets of localization microscopy research. Standardised descriptions of the workflow steps with an automated execution of the workflow, based on extended image analysis programs is achieved by a workflow management system (WfMS).

1 Introduction

Light microscopy is a routine imaging technique in biological and medical research and diagnosis. Although nowadays instrumentation has made substantial progress concerning imaging quality and speed, there has been a gap in resolution between light microscopy (~200 nm) and electron microscopy (~20 nm). This missing scale range has, however, opened new insights into the nanocosmos of a cell and its sub-cellular structures [Mül12]. Localization microscopy, being a candidate to fill this gap, is a technique overcoming resolution limits due to diffraction. During the last decade several setups have been developed and used to answer interesting and challenging questions in the field of cellular biology and molecular biomedicine [Cre11], which we will indicate by two examples.

The data, arising from investigations of localization microscopy, will be shortly characterized as well as the resulting requirements for transmission, formatting, storage, meta data definition, and processing tools. Handling and processing of these data requires the use of a logical and physical network on different levels which has been implemented at the participating institutions.

2 Localization Microscopy closes the Gap

For localization microscopy, standard microscopic optics and fast imaging systems are required. The principle of the so far developed techniques depends on optical isolation and separation of individual dye molecules by their spectral signature. The embodiment (SPDM = Spectral Position Determination Microscopy) used in our collaboration makes use of dye molecules for specific labeling of cellular sub-structures that are able to undergo so called "reversible photo-bleaching", reactions or conformation changes of dye molecules which all result in stochastic molecular blinking. Taking a huge time stack of images (~2000 frames) the switch off/on of each molecule can be detected and the molecular coordinates can be determined precisely (in

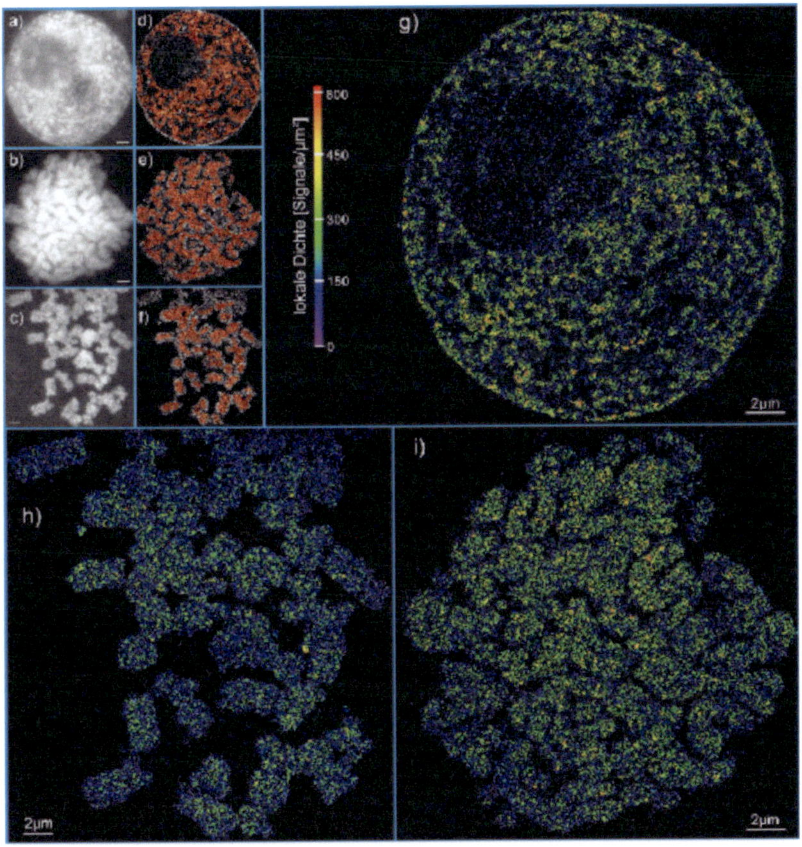

Figure 13: Histone H2B distribution in HeLa cell nuclei and (pro-)metaphase chromosomes [Mül12], showing wide field microscopic images (a–c), individual molecules (red color dots) of merged images from the SPDM time stack (d–f), local density color coding after image evaluation (red color dots) (g–i).

the range of nm). Hence, distances between dye molecules can be calculated in the ten nm range and thus sub-cellular structures can be visualized and measured also in 3D conserved cells or even under vital conditions.

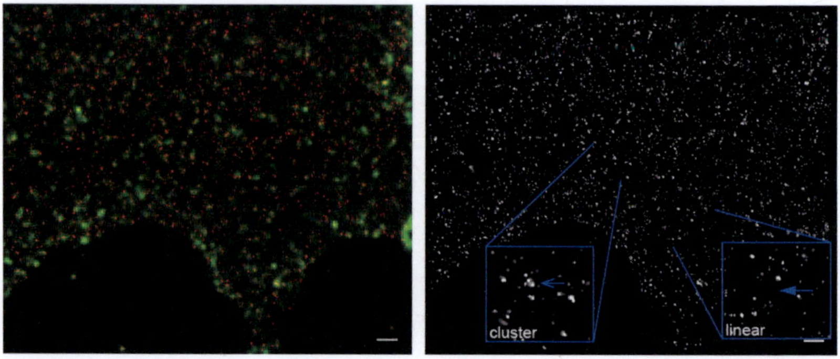

Figure 14: Image section of the membrane of a breast cancer cell after specific labeling of the Her2-receptors by means of fluorescence labeled antibodies. (courtesy J. Neumann, Kirchhoff-Institute for Physics, University of Heidelberg). "cluster" and "linear" refer to different image processing algorithms.

2.1 Examples

In the following two typical examples will be explained: In Figure 13 an example of a cell nucleus and (pro-) metaphase chromosomes are shown. a) - c) show the wide field microscopic images; d) - f) present the merged images from the time stack of SPDM displaying thousands of individual molecules by a color dot. In g) - i) these images are enlarged and coded according to the numbers of next neighbors so that structural information can be elucidated [Boh10]. Such chromatin 2D/3D nano-structures are of importance to understand chromatin rearrangements during repair processes of DNA after exposure to ionizing radiation [Zha15; Fal14a; Fal14b]. This information is used to create and validate a consistent architectural model in the field of radiobiology.

On the left of the Figure 14 an overlay of a standard wide-field image (green) and a localization microscopy image (red points) of a membrane section of a

breast cancer cell is shown. This is where the human epidermal growth factor receptor 2 (Erb B2, a typical breast cancer marker) is specifically labeled by appropriate antibodies. The right image shows the result of localization imaging separately which is obtained from a time series of 1000 image frames (979 x 816 pixels, 150 ms per image). Here, each point represents a single fluorochrome attached to a receptor molecule. The wide-field image, hereby, does not allow the identification of any detailed nano-structural information about the spatial arrangement of the antibodies/receptors [Kau11].
This shortcoming of wide-field image is overcome by the localization image, which reveals details of the formation of receptor clusters or linear arrangements of receptors (inserts) which can be correlated to dimerization induced functional activity [Hau16]. Such analyses help to elucidate mechanisms of breast cancer therapy using antibody treatment (e.g. Herceptin®). These examples indicate the huge progress going along with localization microscopy. However the volume of the data is drastically increasing by orders of magnitudes requiring novel approaches of managing, archiving and analyzing [Pra15].

2.2 Imaging Data and their Requirements

From the examples shown above we assume the digital volume of one cell nucleus of about 20 μm diameter with a resolution of approximately 10 nm is about 32 GB per channel of color. In larger screening experiments the limit of one PB data volume is thus reached easily. For the highly sensitive analyses and structure elucidation, very complex and highly variable algorithms have to be used to avoid artifacts and to find out structural re-arrangements. This includes iterative variation based denoising and deblurring techniques [Asc16].
The algorithms typically follow a given sequence, including background estimation, blob (e.g. cluster) detection, fluorophore location estimation (blob center of mass calculation), followed by postprocessing. In particular, back-

ground estimation is a crucial step, which can be handled via linear or non-linear filtering [D14] where the airy discs from the fluorescent molecules are considered as outliers (e.g. identified by segmentation) and the estimation operates on the remaining data. Yet methods for data fitting with outliers using the ransac algorithm work as well. Further, there are several techniques for estimating the fluorophore location [Mor10; Hua13; RNS15; SS14; TLW02] which differ in complexity.

Postprocessing typically relates the obtained point pattern with the research question. This could include point analysis techniques like distance distribution statistics, clustering, or pattern recognition. One can even consider the image as highly noisy and perform reconstruction using denoising methods such as those based on dictionary learning. For reconstructing a continuous image from point-samples from localization microscopy, a patch-based algorithm with overlapping patches is used. Hereby, the patches are represented by a sparse linear combination of dictionary elements, whereby the dictionary itself is learned from example images using a technique derived from [M10]. The final image is reconstructed by averaging over the overlaps of neighboring patches. Figure 15 shows, as an example, an original binary image without any intensity weight as ususally considered in localization microscopy, whereas Figure 16 shows the OMP learned denoised image.

Still the data is saved and worked on in an ad hoc manner, which with serial computation systems leads to extremely long processing times and a limitation of the selectable volume size due to limitations in the computer memory. The data rate created by a localization microscopy device is in the range of up to GB/s depending on the size of the detected region of interest and the dimensionality (2D/3D) required for scientific investigations.

The original algorithms and techniques had been developed for a PC basis without parallelization, due to the systems available at that time. This strongly limited the handling of large data sets as being necessary in biological research and medical diagnosis especially if a serious significance of statistics

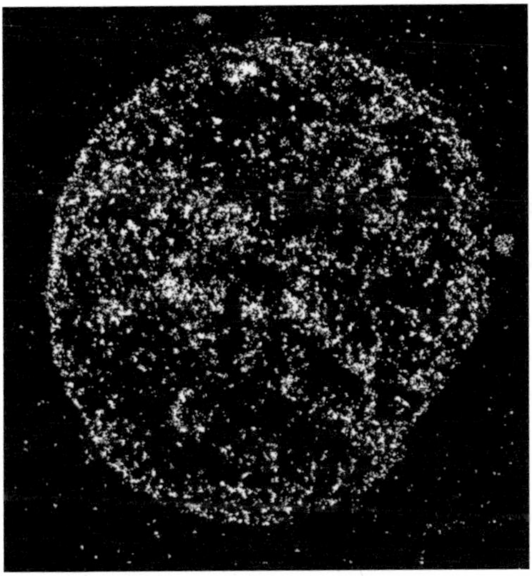

Figure 15: Original binary image.

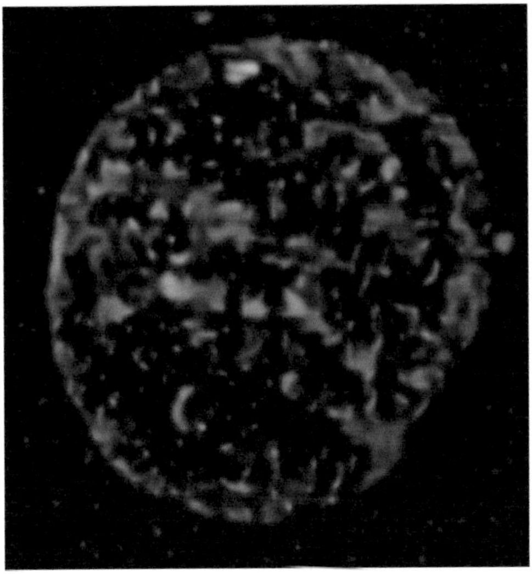

Figure 16: OMP learned denoised image.

is required (i.e. if a large series of cells has to be evaluated). To enhance this process, we developed a pipeline for parallel data analysis.
Variational based methods need, in conjunction with parallel analysis of the data, a synchronous update of all analyzed regions, which required new techniques with message passing. Of course, in addition the access to the data is self-explaining for the user – as best as possible – and fulfills the rules for storage for several years as defined by the DFG funding organization.

3 Localization Microscopy Open Reference Data Repository

Localization microscopy based research can produce datasets of up to 200 TB for many detailed scientific investigations. As this technique is a novel imaging methodology, the archiving, analysis, access and handling of these extremely large datasets necessitates annotating the datasets for experts to share and to compare their findings. Hence, the Localization Microscopy Open Reference Data Repository (LMORDR) has to enable the scientific research community to: Store and access extreme large datasets in a repository, annotate the datasets (enrich the data with new insights), share the annotated datasets (reference data is important for disseminating knowledge), and analyze the datasets interactively. In the Data Life Cycle Lab "Key Technologies" data and microscopy experts jointly developed LMORDR for registering and storing extreme large datasets, a command line client for automatic ingest and access to the extremely large datasets and a web-based tool for executing workflows (with handling of provenance information) and accessing and sharing of the datasets through comprehensive metadata management.

3.1 Architecture

Figure 17 shows the architecture layout required to realize the LMORDR. The principle idea is to harmonize the interaction between various heteroge-

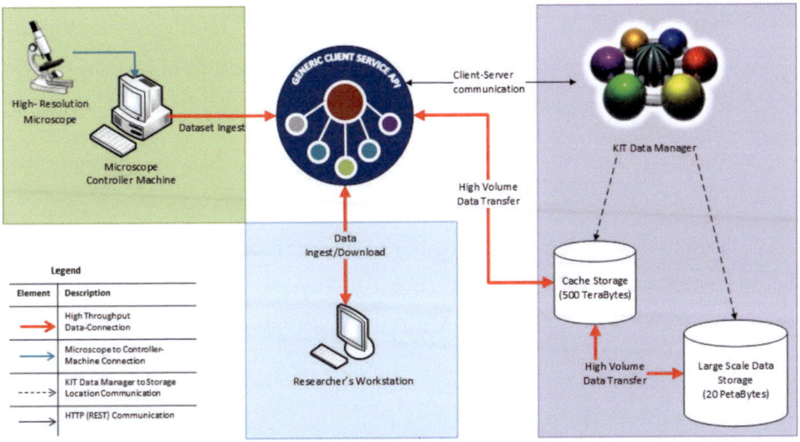

Figure 17: Architecture Layout [Pra15].

neous systems such as the microscope controller machine connected to the high-resolution microscope, researcher's workstations located at different locations, cache storage (required for data-intensive computing), large scale data storage and data repository system (in this case here: KIT Data Manager). To connect such disparate systems, the Generic Client Service (GCS) API is designed and implemented. Furthermore, the GCS API is seamlessly integrated with the KIT Data Manager and the Large Scale Data Storage.

3.2 Base Metadata Model

As the extremely large datasets are difficult to interpret for the researchers, some additional information in the form of metadata needs to be associated with the datasets. A structured metadata model (see Figure 18) based on Core Scientific Metadata Model (CSMD) is established for describing the extremely large datasets of localization microscopy research [Gru14b]. The CSMD consists of three elements:

- Study - The study represents the main topic of the research. Each study is assigned a unique identifier for distinctly identifying each study. Further attributes such as start date and end date of a study, a note describing the detailed description of a study and a manger id for assigning a user to a study can be described.

- Investigation - The investigation represents the subject of each experiment under consideration. To uniquely identify each investigation, a unique identifier is assigned. In the case of localization microscopy research, an investigation topic, e.g. "Investigation on the cell membrane of MCF7 cells" is defined. Multiple investigations can exist for a given study.

- Digital Object - A digital object represents the actual data that is ingested in the Localization Microscopy Reference Data Repository. For sharing and referencing a digital object, a unique digital object identifier is automatically assigned to each digital object.

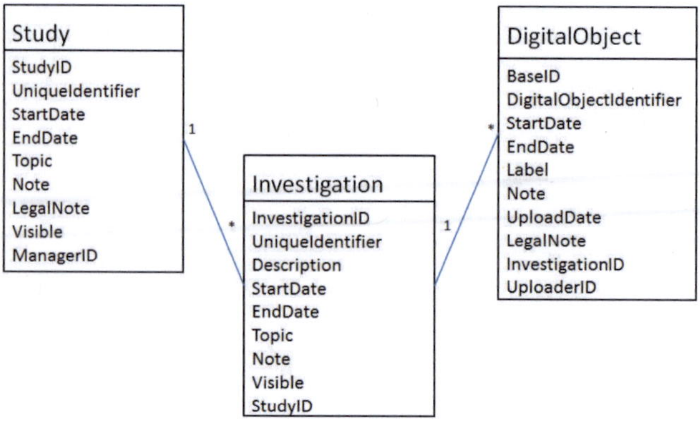

Figure 18: Base Metadata Model.

The base metadata model is only the minimum metadata that is mandatory for storing, sharing and referencing the data in the localization microscopy reference data repository. Moreover, the super-resolution community-specific metadata describing the details of each dataset is maintained by the LMORDR.

The complete community metadata schema [16q] with the necessary services [11e] for handling this metadata are available in the LMORDR.

3.3 Generic Client Service API

The multi-layered GCS API provides a modular solution for the research community to manage the extremely large datasets within the localization microscopy reference data repository. Different components of GCS API (see Figure 19) are briefly explained:

- Access Layer-API (exposes interfaces to connect with various disparate systems), Ingest/Download Workflow component (allows systematic ingest and access of the large datasets),
- Data Transfer Component (provisions bi-directional data transfer between various disparate systems) through high throughput data transfer protocols such as WebDAV,

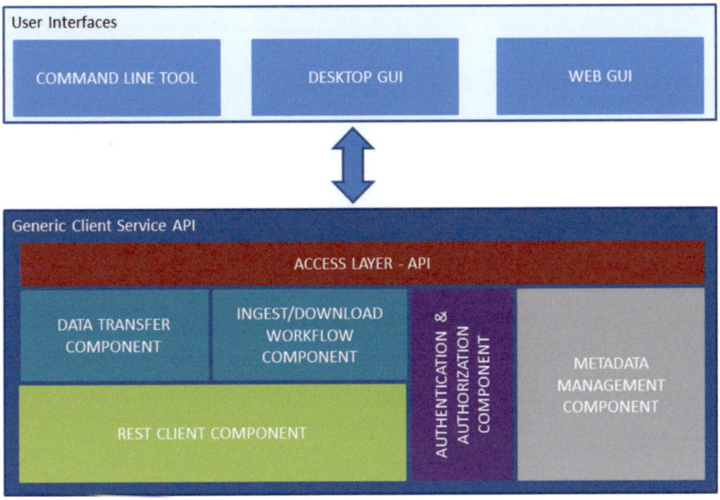

Figure 19: Generic Client Service (GCS) API Architecture.

- REST Client Component (communication with various RESTful services) and Authentication and Authorization Component (a twofold authentication and authorization process,
- OAuth authorization for enabling RESTful services, protocol based authentication and authorization for enabling data transfer and access to data storage),
- Metadata Management Component is responsible for extracting, organizing and modeling the metadata as per the CSMD and the community specific descriptive metadata model. For sharing the localization microscopy metadata, an OAI-PMH [Car02] metadata harvester based on the localization microscopy METS profile [16r] is provided.

4 Workflow and Provenance Management in LMORDR

To generate the result high-resolution images, various scientific workflows are defined by the research community. A workflow for the application of data obtained by super-resolution localization microscopy comprises a systematic organisation of the different image processing algorithms that are necessary to yield the correct results for high-resolution images. To enable a standardised description of the workflow steps with an automated execution of the workflow, a workflow management system (WfMS) is necessary. Not only a WfMS is important but also the provenance information associated with each execution of a workflow is necessary for the applying research community. Comprehensive provenance comprises prospective (workflow definition) and retrospective provenance (workflow execution details) [ZWF06]. Provenance enables to trace the execution of a workflow, analyse the results (intermediate as well as final) produced by the workflow, compare similar workflows and help evolve the existing scientific workflows for producing better results. For completely automating the capturing, modelling (in W3C

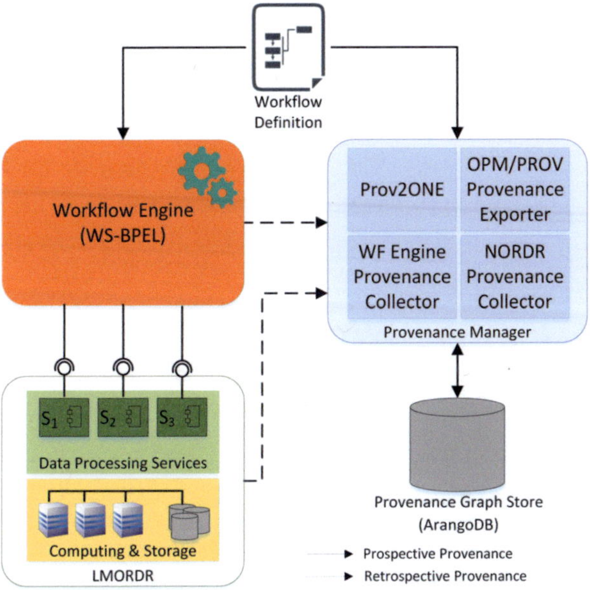

Figure 20: Workflow and provenance management in LMORDR.

ProvONE standard [Cue15]) and storing and querying (graph database) of provenance a dedicated provenance management component on the basis of a workflow management system [13a] is designed, implemented and integrated into LMORDR.

The complete architecture describing the integration of a WfMS and Provenance Manager is shown in Figure 20. The architecture comprises three core components

- Workflow (WF) Engine: The workflow engines interprets the various steps defined in the workflow description language and executes the corresponding image processing algorithms that are deployed as LMORDR Services.
- LMORDR Services: The LMORDR Services is a multi-layered component that offers the various image processing algorithms that are de-

ployed as web services on the high-performance computing cluster for parallel processing and data storage for long-term archival of datasets produced during the workflow execution.

- Provenance Manager: The provenance manager collects, models and stores the entire provenance information generated during the execution of a localization microscopy workflow. The provenance manager comprises four modules:
  - Prov2ONE: This module holds the Prov2ONE [Pra16b] algorithm that is necessary for automatically generating the provenance in W3C ProvONE standard. The ProvONE provenance information is stored as a graph in the ArangoDB [11b] graph database.
  - LMORDR Provenance Collector: The retrospective provenance generated during the execution of the workflow is collected by this component and appended to the ProvONE graph created by the Prov2ONE algorithm.
  - WF Engine Provenance Collector: The additional retrospective provenance collected by the workflow engine is collected by this component and appended to the ProvONE graph created by the Prov2-ONE algorithm. Apache ODE WF engine provides a comprehensive management API [11a] that allows extraction of retrospective provenance.
  - PROV Provenance Exporter: For enabling interoperability among workflow standards (between ProvONE and PROV), this module translates the retrospective provenance from ProvONE to PROV standard.

Finally, the necessary services for storing, querying and analysing the workflow and its associated provenance are provisioned through a REST API [11e].

5 Conclusions and Contributions

Ongoing investigations using localization microscopy in biomedical and cancer research opens an avenue into novel analyses of single molecule arrangements and nanostructures leading to better understanding of molecular mechanisms behind diagnostic outcome and therapy. The acquisition and recording of pointilistic images [Joh99] representing molecular arrangements and dynamics result in huge data sets that have to be managed and archived over long time periods. The DLCL within the initiative LSDMA has made developments towards data management and curation with long term perspectives under the aspects of sustainability and potentials of re-use for different analyses. Here, we have presented the current state of our efforts and investigations in building the comprehensive LMORDR.

Due to the intensive collaboration of computer scientists, biophysicists and biomedical users, constructive approaches towards these aims could be elaborated and successfully implemented and tested. Nevertheless, further studies and developments are necessary to establish a comprehensive system of a user friendly reference data repository and uptodate data management being accepted by the broad community of users in medical research and diagnosis.

Acknowledgments

The authors thank Prof. Dr. Christoph Cremer, Institute for Molecular Biology, Mainz, and Dr. Felix Bestvater, German Cancer Research Center, Heidelberg, for instrumentation support. Furthermore the authors thank LSDMA for financial support. ES and MH are indepted to Prof. Dr.-ing. Prinz zu Pell, University of Laxenburg, Austria, for pulling our thoughtfulness to the bitter pill when polling the community casts a pall over never-to-be-applied approaches.

Quoted LSDMA Publications

[Gru14b] Richard Grunzke et al. "Device-driven metadata management solutions for scientific big data use cases". In: *2014 22nd Euromicro International Conference on Parallel, Distributed, and Network-Based Processing*. IEEE. 2014, pp. 317–321.

[Pra15] Ajinkya Prabhune et al. "An Optimized Generic Client Service API for Managing Large Datasets within a Data Repository". In: *Big Data Computing Service and Applications (BigDataService), 2015 IEEE First International Conference on*. Mar. 2015, pp. 44–51. DOI: 10.1109/BigDataService.2015.25.

[Pra16b] Ajinkya Prabhune et al. "Prov2ONE: An Algorithm for Automatically Constructing ProvONE Provenance Graphs". In: *Provenance and Annotation of Data and Processes: 6th International Provenance and Annotation Workshop, IPAW 2016, McLean, VA, USA, June 7-8, 2016, Proceedings*. Springer International Publishing, 2016, pp. 204–208. ISBN: 978-3-319-40593-3. DOI: 10.1007/978-3-319-40593-3_22.

Other References

[11a] *Apache ODE Management API*. 2011. URL: http://ode.apache.org/management-api.html.

[11b] *Arango DB*. 2011. URL: https://www.arangodb.com/.

[11e] *Localiation Microscopy — Swagger UI*. 2011. URL: http://datamanager.kit.edu/masi/localizationmicroscopy/swagger-ui/%5C#/.

[13a] *Apache ODE*. 2013. URL: http://ode.apache.org/.

[16q] *Localiation Microscopy Meta Data Scheme*. Mar. 2016. URL: http://datamanager.kit.edu/masi/localizationmicroscopy/2016-03/LocalizationMicroscopy.xsd.

[16r] *Localiation Microscopy — Nanoscopy METS Profile*. 2016. URL: http://datamanager.kit.edu/masi/localizationmicroscopy/mets/nanoscopy-METS-profile.xml.

[Asc16] K.P. Aschenbrenner et al. "Compressed sensing denoising for segmentation of localization microscopy data". In: *IEEE Int. Conf. Comput. Intelligence Bioinf. Comput. Biol. (CIBCB 2016, 5.-7.10.2016)* (16), in press.

[Boh10] M. Bohn et al. "Localization microscopy reveals expression-dependent parameters of chromatin nanostructure". In: *Biophysical Journal* 99 (2010), pp. 1358–1367.

[Car02] L. Carl et al. *The Open Archives Initiative Protocol for Metadata Harvesting, Version 2.0 [Online]*. 2002. URL: http://www.openarchives.org/OAI/2.0/openarchivesprotocol.htm.

[Cre11] C. Cremer et al. "Superresolution imaging of biological nanostructures by spectral precision distance microscopy". In: *Biotechnology Journal* 6 (2011), pp. 1037–1051.

[Cue15] V. Cuevas-Vicenttín et al. "A PROV Extension Data Model for Scientific Workflow Provenance". In: *Private communication* (2015).

[D14] Evanko D. "Taming the image background beast". In: *Nature Methods* 228 (2014), p. 11.

[Fal14a] M. Falk et al. "Giving OMICS spatiotemporal dimensions by challenging microscopy: From functional networks to structural organization of cell nuclei elucidating mechanisms of complex radiation damage response and chromatin repair – PART A (Radiomics)". In: *Crit. Rev. Eukaryot. Gene Express.* 24 (2014), pp. 205–223.

[Fal14b] M. Falk et al. "Giving OMICS spatiotemporal dimensions by challenging microscopy: From functional networks to structural organization of cell nuclei elucidating mechanisms of complex radiation damage response and chromatin repair - PART B (Structuromics)". In: *Crit. Rev. Eukaryot. Gene Express.* 24 (2014), pp. 225–247.

[Hau16] M. Hausmann et al. "Challenges for super-resolution microscopy and fluorescent nano-probing: Understanding mechanisms behind tumour development and treatment". In: *Int J Cancer* (2016), submitted.

[Hua13] F. Huang et al. "Video-rate nanoscopy using sCMOS camera-specific single-molecule localization algorithms". In: *Nature Methods* 10 (7 2013), pp. 653–658.

[Joh99] Gage John. "Seurat's Silence". In: *Color and Meaning: Art, Science, and Symbolism*. University of California Press, 1999, pp. 223–225. ISBN: 9780520226111.

[Kau11] R. Kaufmann et al. "Analysis of Her2/neu membrane protein cluster in different types of breast cancer cells using localization microscopy". In: *Journal of Microscopy* 242 (2011), pp. 46–54.

[M10] Elad M. *Sparse and redundant representations: From theory to applications in signal and image processing, Haifa, Israel*. Springer Science+Business Media, New York, USA, 2010.

[Mor10] K.I. Mortensen et al. "Optimized localization analysis for single-molecule tracking and super-resolution microscopy". In: *Nature Methods* 7 (5 2010), pp. 377–381.

[Mül12] P. Müller et al. "Analysis of fluorescent nanostructures in biological systems by means of Spectral Position Determination Microscopy (SPDM)". In: *Current microscopy contributions to advances in science and technology*. Ed. by A. Méndez-Vilas. Vol. 1. 2012, pp. 3–12.

[RNS15] B. Rieger, R.P.J. Nieuwenhuizen, and S. Stallinga. "Image processing and Analysis for Single-Molecule Localization Microscopy: Computation for nanoscale imaging". In: *IEEE signal Processing Magazine* 32 (2015), pp. 49–57.

[SS14] A. Small and S. Stahlheber. "Fluorophore localization algorithms for super-resolution microscopy". In: *Nature Methods* 11 (3 2014), pp. 267–279.

[TLW02] R.E. Thompson, D.R. Larson, and W.W. Webb. "Precise nanometer localization analysis for individual fluorescent probes". In: *Biophysical Journal* 82 (5 2002), pp. 2775–2783.

[Zha15] Y. Zhang et al. "Radiation induced chromatin conformation changes analysed by fluorescent localization microscopy, statistical physics, and graph theory". In: *PLoS ONE* 10 (2015), e0128555. DOI: 10.1371/journal.prone.0128555.

[ZWF06] Y. Zhao, M. Wilde, and I. Foster. "Applying the virtual data provenance model". In: *Proceedings of the 2016 International Provenance and Annotation Workshop, Chicago USA* (2006), pp. 148–161.

Design and Setup of Experiment Specific Data Analysis and Storage Systems for Photon Science and Heavy Ion Physics

Martin Gasthuber[a], Kilian Schwarz[b]

[a] Deutsches Elektronen-Synchrotron DESY, Hamburg
[b] GSI Helmholtzzentrum für Schwerionenforschung GmbH, Darmstadt

Abstract The LSDMA DLCL Structure of Matter comprises the projects European XFEL and PETRA III at DESY in Hamburg as well as FAIR (Facility for Antiproton and Ion Research) at GSI in Darmstadt. Within the context of the FAIR project, significant improvements in the XrootD storage infrastructure have been developed and deployed at the ALICE Tier 2 centre at GSI. Within the context of EuXFEL and PETRA III, a new storage & analysis system has been developed. In the case of Petra III, the system is in operation since 2015 – the EuXFEL system is in early pre-production phase.

1 Introduction

1.1 Facility for Antiproton and Ion Research (FAIR)

The new international research center FAIR (Facility for Antiproton and Ion Research [10a]) is currently under construction next to GSI in Darmstadt. FAIR will support a wide variety of science cases: extreme states of matter using heavy ions (CBM), nuclear structure and astrophysics (NUSTAR), hadron physics with antiprotons (PANDA), atomic and plasma physics, as well as biological and material sciences (APPA). The high beam intensities at FAIR constitute various challenges; especially in collaborative computing. Full operation of the Modular Start Version is foreseen for the first half of the next decade. The major experiments at FAIR will implement a novel triggerless detector read-out, without conventional first-level hardware triggers, relying exclusively on software-based event filters. The traditional separation between data acquisition, trigger, and off-line processing is merging into a single, hierarchical data processing system, handling a data stream from the detectors exceeding 1 TB/sec. To cope with the enormous computing challenges, several sites in the surrounding of GSI [69] will be connected with a high-speed Metropolitan Area Network via fibre link allowing the off-loading of processing between the sites. That combined Tier 0/1 system will be integrated in an international grid/cloud infrastructure. A prototype named PandaGrid exists already for the PANDA experiment using the AliEn [Sai03] middleware, originally developed by the ALICE [93] Collaboration. The Grid monitoring and data supervision are done via MonAlisa. The basis software framework for simulation, reconstruction, and data analysis is FairRoot on top of which the FAIR experiments develop experiment specific software. In order to meet peak demands for computing, it may be necessary to offload some of the computing tasks to public or community clouds. The resource requirements are dominated by CBM and PANDA. Current estimates for the sum of all experiments are 200.000 cores and 30 PB storage space for the first year of data taking.

1.2 PETRA III

The PETRA III [17b] storage ring went into operation in 2009 and is the world's most brilliant storage-ring-based X-ray radiation source. It features 14 experimental stations with up to 30 instruments. The recent completed upgrade to 24 experimental stations will significantly enlarge the rate and volume of data to be processed and stored. PETRA III provides excellent opportunities in the fields of material research and molecular biology. Its tightly collimated beam with short wavelength allows to investigate small samples and to resolve for example the complex structure of ribosomes.

1.3 European XFEL

At Hamburg, a new research facility for a free electron laser – the European XFEL [17c] – is being built. The worldwide unique source of X-rays will provide ultrashort X-ray flashes (27000 times per second) with a brilliance billion times higher than that of conventional X-ray sources. The radiation has properties similar to LASER light. The XFEL is expected to go into operation in 2017. At present twelve countries participate in this international project with DESY being the main shareholder. The XFEL accelerates bunches of electrons to high energies. These electrons are then passed through specially arranged magnets, so-called undulators, producing the X-ray flashes. Many research fields will benefit from this new radiation source. It allows to resolve atomic details of viruses, to decipher the molecular composition of cells, to film chemical reactions, or to study processes such as those occurring deep inside planets.

2 DLCL Structure of Matter: FAIR

Within the context of the FAIR project, significant improvements in the XrootD [12] storage infrastructure have been developed and deployed at the ALICE Tier 2 centre at GSI.

2.1 The ALICE Tier 2 Centre and the National Analysis Facility for ALICE at GSI

The ALICE Tier 2 centre and the National Analysis Facility at GSI provide a computing infrastructure for ALICE Grid and for local use of the German ALICE groups.

Over the years GSI participates in centrally managed ALICE Grid productions and data analysis activities, as well as in data analysis of individual users submitting their jobs to the ALICE Tier 2 centre.
In 2015, 7.5% of all successfully computed ALICE grid jobs have been running at the two German grid sites, the GSI Tier 2 centre, and Forschungszentrum Karlsruhe (ALICE Tier 1 centre). This corresponds to the pledged CPU resources for 2015: 13400 HEP-SPEC06 for GSI Tier 2 and 30000 HEP-SPEC06 for FZK.
The storage resources pledged at GSI to the global ALICE community (1700 TB) are provided via a Grid Storage Element which consists of a set of XrootD daemons running on top of a Lustre [03] file system. The XrootD setup of the ALICE Tier 2 centre is moreover being used as test bed for new developments providing I/O optimisation and integration in local HPC environments.

2.2 The XrootD Forward Proxy

The main elements of the GSI Storage Element are the three XrootD data servers and the XrootD redirector. The latter uses XrootD's *split directive* in order to redirect clients from world-routable IPs to the external interfaces of the XrootD data server machines and clients with private IPs to the internal interfaces, profiting from the high bandwidth of the local InfiniBand fabric.
HPC clusters are often used in an isolated environment where direct connections between the cluster's worker nodes and the internet are partially or fully restricted. An XrootD proxy enables the site admin to allow worker nodes to read input files from and write output files to remote sites while adhering to the aforementioned restriction. Such a setup is currently in production at GSI.
In order to increase availability and decrease error rate in case of a proxy outage, it is planned to introduce a second XrootD proxy server as well as a pair of XrootD proxy redirectors, which will constitute a HA setup where

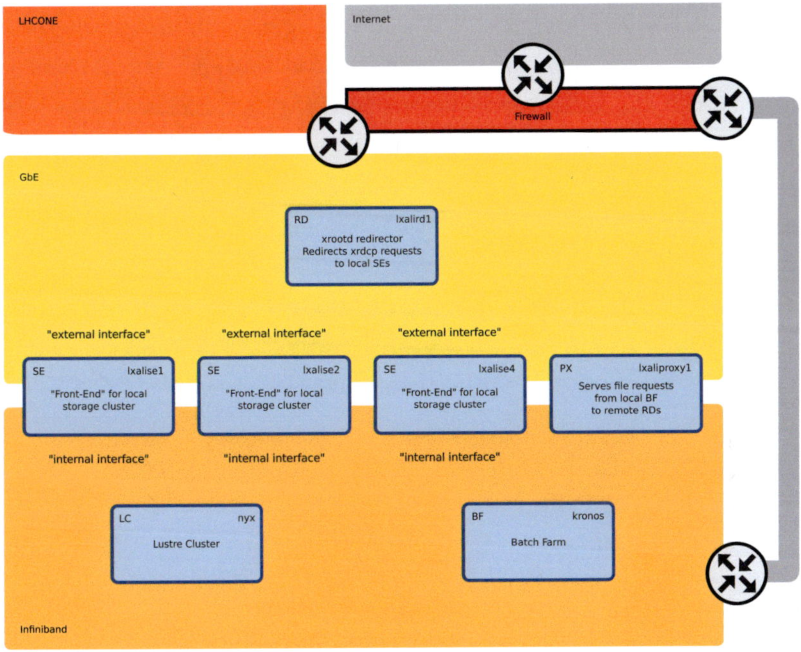

Figure 21: GSI's ALICE Tier 2 setup including XrootD forward proxy.

for every machine type one is allowed to be down at any given time without losing availability of the whole system.
The experience gained in the context of operating the ALICE Tier 2 centre and the National Analysis Facility as a production service for ALICE serves as a guideline for a distributed computing environment for FAIR.

2.3 Optimising I/O Performance by using an XrootD Plug-in

Granting access to scientific data that is already available on GSI's Lustre filesystem via XrootD at the Alice Tier 2 centre revealed infrastructure-related I/O bottlenecks. This section describes the proposed solution, an XrootD plug-in, and successfully performed tests while using it.

Via the ALICE Tier 2 centre at GSI, a computing infrastructure is being provided for granting access to remote and local storage to the ALICE community. In ALICE's AliEn grid framework specific data is accessed through an XrootD infrastructure. XrootD enables the use of of this data through a scalable federated storage system. At the local GSI HPC infrastructure, however, all access to storage, including the 1700TB storage resources pledged to ALICE, is provided by a Lustre filesystem. While XrootD provides good scalability, Lustre already provides low latency and high I/O bandwidth to locally available data. For this reason it has been decided that XrootD will serve data by reading from Lustre instead of providing extra storage for XrootD data servers.

Grid jobs requiring the same data are often scheduled on the same site to lower the need for traffic between sites. Additionally, the German ALICE group at GSI often reuses data many times after it has been copied to GSI storage once. This means that data stored at GSI remains for quite some time on site and it is essential to optimize the I/O performance.

With the current storage infrastructure at GSI, namely the access to Lustre through the XrootD data servers, the following room for improvement has been identified:

1. The current XrootD data servers can provide only limited I/O bandwidth
2. All data read locally from Lustre needs to be sent over the Network twice (Lustre to XrootD data server & XrootD server to client), effectively doubling the network traffic for an I/O operation

All clients using XrootD to access ALICE grid data request it through XrootD data servers, this means that I/O traffic needs to go through the limited link one data server can provide and that Lustre's full I/O speed cannot be utilized directly.

The proposed solution is to use the XrootD client plug-in API to redirect underlying access to data on Lustre directly, bypassing the XrootD workflow

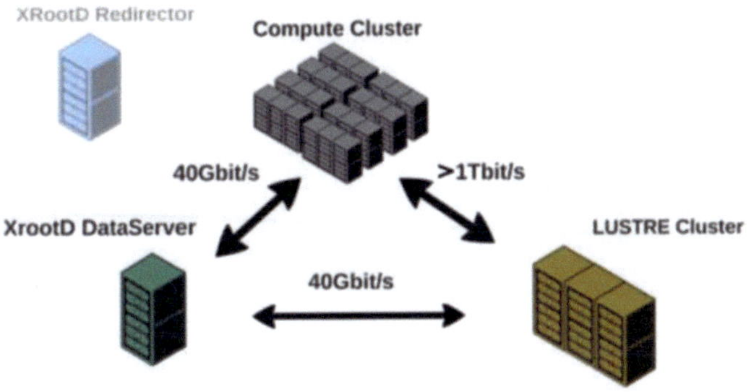

Figure 22: The XrootD storage access at GSI.

if the data is locally available. The XrootD client plug-in API comes with the advantage to change XrootD's underlying I/O operations, so that higher level software (e.g. ROOT, xrdcp, ...) can use the plug-in without the need to know of its existence.

Algorithm 1 A simplified example of the plug-in's "Open" call.

```
virtual XrootDStatus Open(params param){
{
    // use local file implementation
    // if local file is available
    if (DataisLocal(param.path)) {
      XRootDStatus* return_stat;
      file->open(param);
      if(file->fail())
        return_stat=new XRootDStatus(XrdCl::stError);
      else
        return_stat=new XRootDStatus(XrdCl::stOK);
      handler->HandleResponse(ret_st,0);
      return *return_stat;
    }
    //use the XrootD standard implementation if not
    return xfile.Open(param);
}
```

Tests using the current plug-in show promising results, granting clients access to Lustre with comparable I/O speeds without the need to use the full XrootD workflow.

In conclusion, an XrootD client plug-in has been implemented, redirecting client I/O to data on Lustre, effectively improving the I/O performance by bypassing the need to read indirectly via the XrootD data server. The current tests show that such a plug-in can be used to adjust XrootD to specific needs as described above.

3 European XFEL and PETRA III

3.1 Introduction and Challenge

The development of detectors at 3rd generation light sources are currently outpacing experimental methods and data acquisition. Single clients will produce 0.5 GBytes/sec and the next generation is already pushing for 6 GBytes/sec. For 30 beamlines the expected averaged aggregated rate is of 50 to 80 GBytes/sec, depending on detector deployments. Also, measurements last from a few hours to a few days resulting in many single data sets up to tens of TBs each. From next generation detectors we also expect multi GBytes/sec spread over many 10GE connections. Furthermore, there is a very dynamic experimental setup with inherent burst nature and a very heterogeneous environment regarding technology, social context and requirements. In order to support better data control and shorter turnaround cycles, the new system has to allow high speed data access within seconds after data have been generated by the detector, within a few minutes for full scale data analysis using multiple CPUs and within hours to be archived to tape media and to be available for external (remote) access.

This article presents the selected components, the overall architecture and experiences with our new architecture and services deployed at the local Petra III facility being in full production since April 2015. Technology choices

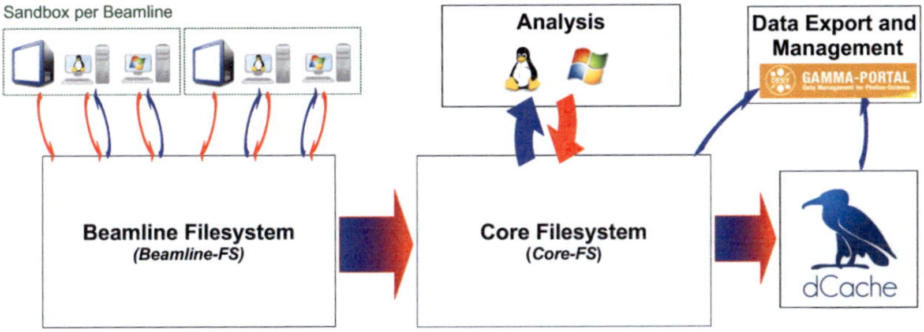

Figure 23: Overall architecture and data flows.

were undertaking over a period of 10 months. The work involved a close collaboration between central DESY-IT, beamline controls, and beamline support staff. Our approach integrates leading edge HPC technologies for storage systems and protocols. In particular, our solution uses a multi filesystem instance with multi protocol data access, while operating within a single namespace and using automated data management behind the scenes, for archive and data export purposes.

3.2 Industry Cooperation

A cooperation on core technical and technological areas was established between DESY-IT and IBM to include industrial research and development experiences and skills. DESY and IBM started a cooperation to develop and build a system based on the General Parallel Filesystem [98] (GPFS) technology from IBM. IBM internally sponsors the internal activities for Beta-Hardware and, more important, the work time spent from various teams in research and development groups.

3.3 New System for Petra III

The picture (fig. 23) shows the overall composition and the automated data flows (information life cycle). The initial architecture and setup discussed at [Str15]. Together with the transactional characteristics for starting and ending a new experiment, it covers the whole life cycle for experimental data taking and analysis. During data taking, the controlled access is through NFS, SMB and a specialized ZMQ based channel (see section below about online data analysis). The initial physical media for taken data are SSD based storage pools, allowing high bandwidth and low latency access. Getting matured, data migrates to spinning disk based resources, still residing in the Beamline-FS specially configured for that purpose (access rights and modes, IO pattern and access protocols). After a few minutes of data aging, the next automated process generates copies to the Core-FS, built with a different configuration to support full authentication, access-control-list, high availability etc. Once the experiment concludes and all data from Beamline-FS is proved to be copied over to the Core-FS, all resources at the Beamline-FS will be removed, thus being ready for the next experiment at the same experimental station. As soon as the data arrives in the Core-FS, it can be accessed (in very high speed) in parallel, from the Analysis-Cluster (Windows & Linux platform), selected datasets can be automatically copied to the tape-archive (using the existing site installations of a dCache system and preserving ACLs) and made available for external access through http (browser based access) and ftp. An additional benefit of using dCache as an archive is the ability to allow seamless and controlled data access through NFS to the archived data to/from any host on site.

3.4 Initial Deployment and further Developments

The initial deployment of the new system has been done for the onsite Petra III light source experiments and is used in full production since April 2015. Since then, in depth discussions and tests have been done by the local EUXFEL computing group to verify the applicability of that architecture as a blueprint for the EUXFEL data and analysis system. Initial deployments for the EUXFEL system have already been started and will be the main focus in the following months. Both Petra III and EUXFEL will demand for

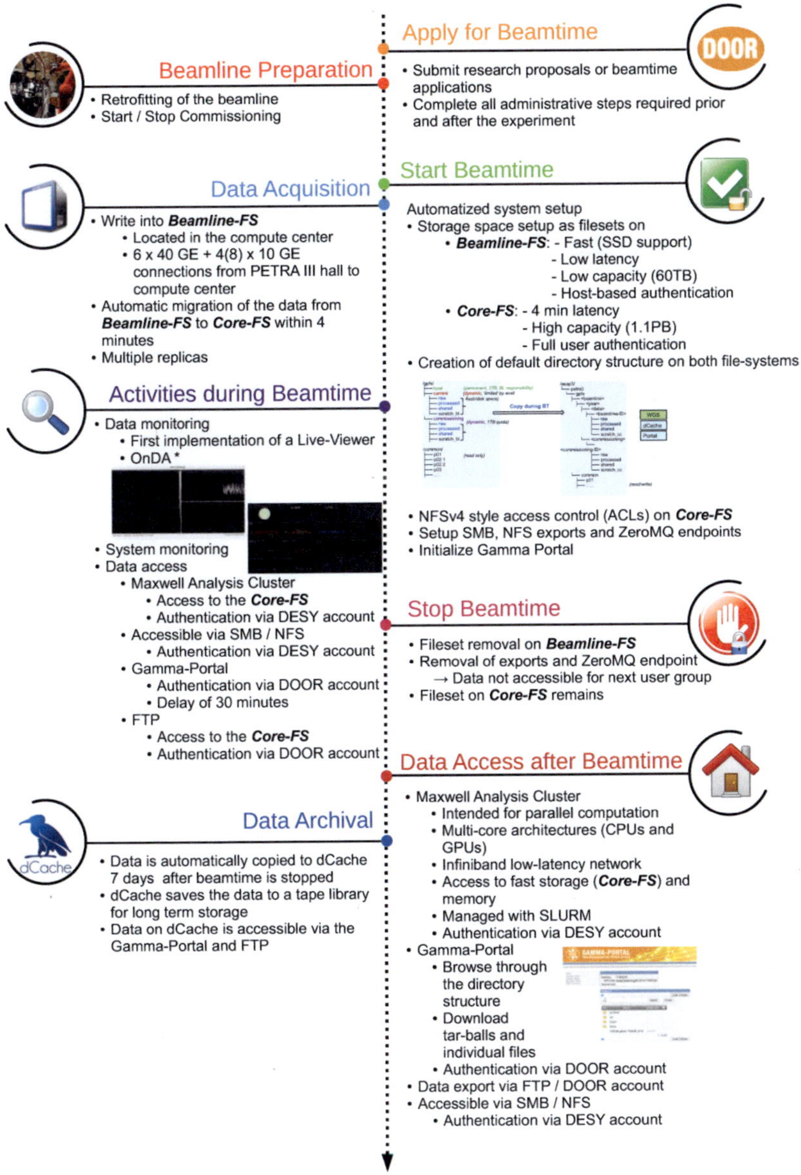

Figure 24: Operations & Timeline from User Perspective.

even higher data rates and volumes compared to existing experimental setups and detector technologies. The IO-profile is expected to change with higher demands in burst capturing and recursive selected reads while data analyses runs. The picture (Figure 24) shows the the typical steps from user perspective including the activity outline from 'behind the scene'.

3.5 Near Realtime Data Access – Online Data Analysis

Next generation of experiments will require controlled and fast access (bandwidth and latency) to the most current generated data to allow immediate experiment control. Local scientist have developed experimental setups where samples are constantly flowing in a liquid or gaseous jet across a pulsed X-ray source which has a repetition rate of up to 120 Hz. Significant amounts of sample are consumed in a very short time, and the data generated by the

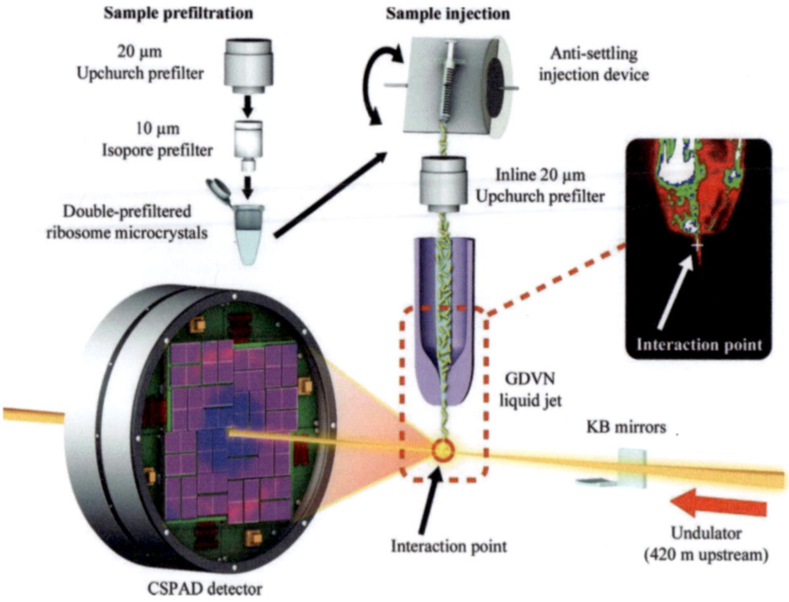

Figure 25: Next generation experiment setup.

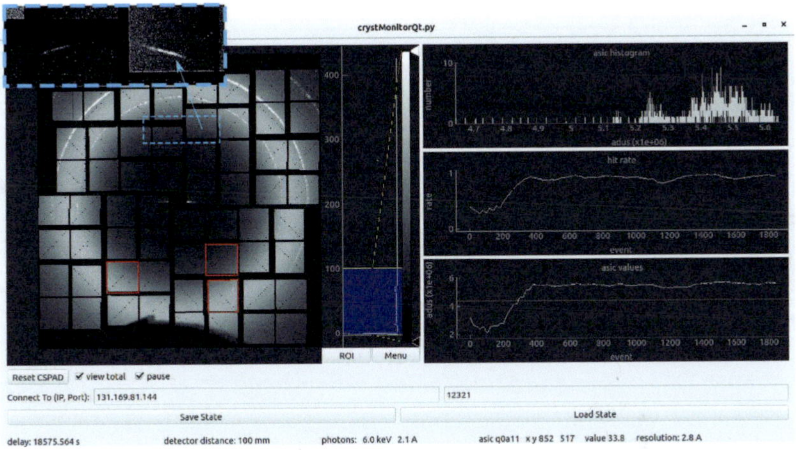

Figure 26: ONDA - Live-View.

instruments requires a large amount of storage space. Furthermore, experimental parameters, such as the degree of molecular alignment in controlled imaging experiments, or the hit rate and resolution in an SFX[1] experiment, must be kept within acceptable bounds. By monitoring experimental conditions in close to real time, the experiment may be maintained in optimal alignment, or alternatively, one may pause the experiment to correct unfavorable conditions, thereby preventing the collection of unfavorable data while preserving valuable sample. Figure 25 shows an example of the primary components and configuration of such experiment setups currently in preparation. Overlooking instrument development for the future (i.e. higher pulse repetition rates,...) there is no way around real-time analysis and data reduction. OnDA (Online Data Analysis) (Figure 26) is a fast online feedback framework which provides the possibility to decide in near real-time about the quality of the data produced in serial X-ray diffraction and scattering experiments [Mar16]. It is designed on a highly modular basis and provides stable and efficient real-time monitors for most common types of experiments. Re-

[1] Serial femtosecond X-ray crystallography

cent beamtimes completed at the local Petra III facility, shows 'better than expected' integration with the new storage system supporting all required criteria. This integration work is ongoing, expecting more experiments with similar demands. Building a generic solution, supporting all types of data flow control and dispatch, meeting all performance criteria, is the base goal for the ongoing development effort. The local developed HiDRA software package introduced a generic layer to allow a flexible data flow configuration between detector and the first 'touch down' of the data in the GPFS system for any type of online *synchronous* data analysis. Further details of HiDRA are shown in the section *Data Transfer and Online Analysis of Scientific Data* of the article *Performance and Power Optimization* in this book.

3.6 Conclusion and Contributions

The new system fulfils all requirements and shows very good adaptability for new use-cases being implemented recently. The architecture shows good scaling (in terms of bandwidth and latency) which will meet the experimental conditions for the next few generation of experimental setups. Work has already been started, with promising initial results, to fully support any type of 'online data analysis', leaving the data in 'flight', control the flow direction and quality, before 'landing' at the storage layer. Detailed, more technical, information could be found in [Die15], [Die16] and in the system documentation for the Petra III system at [Tea16].

Quoted LSDMA Publications

[Die15] Stefan Dietrich et al. *ASAP3: New Data Taking and Analysis Infrastructure for PETRA III.* HEPiX Workshop, Univ. Oxford/UK. Deutsches Elektronen Synchrotron DESY, 2015. URL: https://indico.cern.ch/event/346931/contributions/817807/attachments/684652/940445/Dietrich%5C_ASAP3%5C_new%5C_data%5C_taking%5C_infrastructure.pdf.

[Die16] Stefan Dietrich et al. *ASAP3: Status Update and Activities for XFEL.* HEPiX Workshop, DESY/Zeuthen. Deutsches Elektronen Synchrotron DESY, 2016. URL: https://indico.cern.ch/event/466991/contributions/1143592/attachments/1260614/1862916/Dietrich%5C_ASAP3%5C_Status%5C_Update%5C_and%5C_XFEL%5C_Activities.pdf.

[Str15] M. Strutz et al. "ASAP3 - New Data Taking and Analysis Infrastructure for PETRA III". In: *J. Phys. Conf. Ser.* 664.4 (2015), p. 042053. DOI: 10.1088/1742-6596/664/4/042053.

Other References

[03] *Lustre*. 2003. URL: http://lustre.org/.

[10a] *FAIR*. 2010. URL: https://www.gsi.de/en/researchaccelerators/fair.htm.

[12] *xrootd*. 2012. URL: http://xrootd.org/.

[17b] *Petra III, Synchrotron Radiation Source at DESY*. Deutsches Elektronen Synchrotron DESY. Notkestrasse 85, 22607 Hamburg, Germany, 2017. URL: http://photon-science.desy.de/facilities/petra%5C_iii/index%5C_eng.html.

[17c] *XFEL*. 2017. URL: http://www.xfel.eu/.

[69] *GSI*. 1969. URL: https://www.gsi.de/en/start/news.htm.

[93] *ALICE*. 1993. URL: https://home.cern/about/experiments/alice.

[98] *GPFS, General Parallel File System*. 1998. URL: https://en.wikipedia.org/wiki/IBM%5C_General%5C_Parallel%5C_File%5C_System.

[Mar16] Valerio Mariani et al. "OnDA: online data analysis and feedback for serial X-ray imaging". In: *Journal of Applied Crystallography* 49.3 (June 2016), pp. 1073–1080. DOI: 10.1107/S1600576716007469.

[Sai03] P. Saiz et al. "AliEn – ALICE environment on the GRID". In: *Nucl. Instrum. Meth.* A502 (2003), pp. 437–440. DOI: 10.1088/1742-6596/119/6/062012.

[Tea16] DESY ASAP Team. *ASAP3: System Documentation*. System Documentation. Deutsches Elektronen Synchrotron DESY, 2016. URL: https://confluence.desy.de/display/ASAP3/ASAP3++Data+Storage+for+PETRA+III.

UNICORE-based Neuroimaging and Data Sharing Workflow

André Giesler[a]

[a] Forschungszentrum Jülich, Jülich Supercomputing Centre, Jülich

Abstract We contributed to the realization of a complex image processing workflow for reconstructing the three-dimensional nerve fibers of postmortem brains by using the Polarized Light Imaging technique. Images of brain slices are processed with a chain of tools that have been integrated in a UNICORE-based workflow exploiting many of its features, such as automated processing, control structures, and data sharing. The introduction of the UNICORE workflow approach for this particular use case led to several benefits by enabling a time-saving automated processing, achieving better reproducibility, and performing routine data production for scientists without knowing the complex interaction of supercomputing and data infrastucture.

1 Introduction

The Health Data Life Cycle Lab concentrated its activities on the domain of Three-dimensional Polarized Light Imaging (3D-PLI). This neuroimaging technique is used at the Institute of Neuroscience and Medicine (INM-1), Forschungszentrum Juelich, to reconstruct the three-dimensional nerve fiber architecture in postmortem mouse, rat, and human brains at the micrometer scale [Axe11]. The examination of a human brain with 3D-PLI generates about 2,500 histological sections, which are digitized at 1.3 μm pixel size resulting into image sizes per section of about 70,000 x 100,000 pixel. The subsequent post processing and the extraction of fiber orientations from the microscope images requires a complex chain of tools. These tools have been integrated in a UNICORE workflow towards a fully automated and parallelized image processing utilizing advanced supercomputing infrastructure efficiently.

2 Methods

The supercomputing facilities at the Juelich Supercomputing Centre (JSC) served as the infrastructure to carry out the PLI workflow studies. In particular, the GPU-Cluster JuDGE (Juelich Dedicated GPU Environment) [11d] and the JuRECA cluster (Juelich Research on Exascale Cluster Architectures) [15c] proved perfect for setting up a streamlined 3D-PLI analysis utilizing the benefits from case-sensitive GPU and CPU acceleration. The Grid middleware UNICORE (Uniform Interface to Computing Resources) [Str10] and its incorporated workflow engine were chosen to create an easy-to-use workflow of the PLI processing pipeline. UNICORE is an open source middleware that facilitates access to supercomputing resources. It offers integrated workflow management support, controls the execution of sequential and parallel compute jobs at one or multiple sites, and makes data resources available in a seamless and secure way. A full-featured graphical workflow editing and runtime monitoring is part of the UNICORE Rich Client (URC) [Dem10].The 3D-PLI workflow comprised image calibration, independent component analysis, image segmentation, stitching, and fiber orientation determination as described in [Axe11]. Figure 27 shows a high level view of the sequential processing of the PLI workflow tools. Individual parallelization was realized at the level of the implemented algorithms, the processing strategies, and the workflow itself. The amount of data for a single brain section is in the order of magnitude of up to 750 GB, with intermediate results at the same scale. Thus, the total amount of processed data easily adds up to several TB of data movement within a typical 3D-PLI workflow.

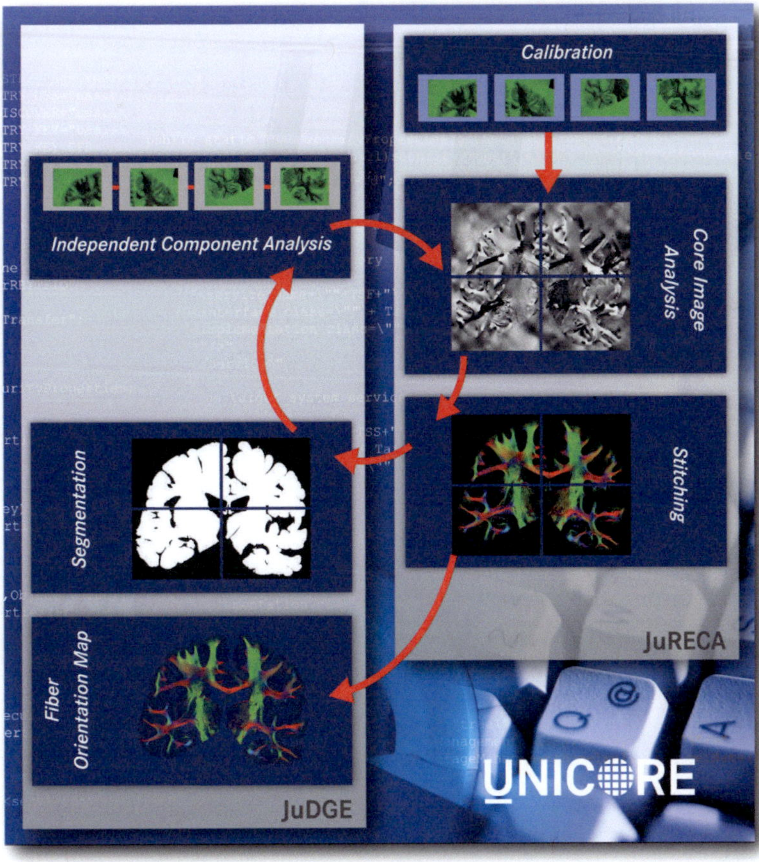

Figure 27: Image processing of the 3D-PLI Workflow.

3 Results

The deployment of the UNICORE-based 3D-PLI workflow regarding an individual 1.3 μm-resolution human brain section that spans an area of 70,000 x 100,000 pixel reduced significantly processing time from days to hours, which is a relevant factor considering the thousands of sections to be analyzed in a whole human brain study. UNICORE turned out to be a valuable tool serving both the software developer by integrating their image processing tools as well as the scientific user lacking in deep knowledge of how to use a supercomputer infrastructure. Untrained neuroscientists were able to perform complex data analysis and routine data production without knowing all details about the different data inputs, calls and requirements of individual software packages of the workflow. This setup clearly minimized operation failures as compared to manual processing of individual software packages. Furthermore, the workflow could be used as performance measurement tool for the utilized supercomputers. In order to take advantage of specific features of different supercomputers (e.g., GPU vs. faster CPU), the compute performance of both systems JuDGE and JuRECA was addressed by the workflow. For the segmentation, stitching, and the fiber orientation software, it turned out to be a performance gain to utilize the faster CPUs of the JuRECA system. Figure 28 shows a resulting fiber orientation map produced by the 3D-PLI workflow.

Current activities of the DLCL Health are focused on the aspect of workflow provenance. Besides the need to facilitate computations and experiments by making use of scientifc workflows, the availability of provenance information is as important as the results of the scientific analysis itself. Scientific users may want to track the origin of a data product in order to verify its conformity or to check the used software versions, hardware environment or contributors. For that reason the DLCL Health intensified its activities in that field by developing the UniProv provenance management system which captures

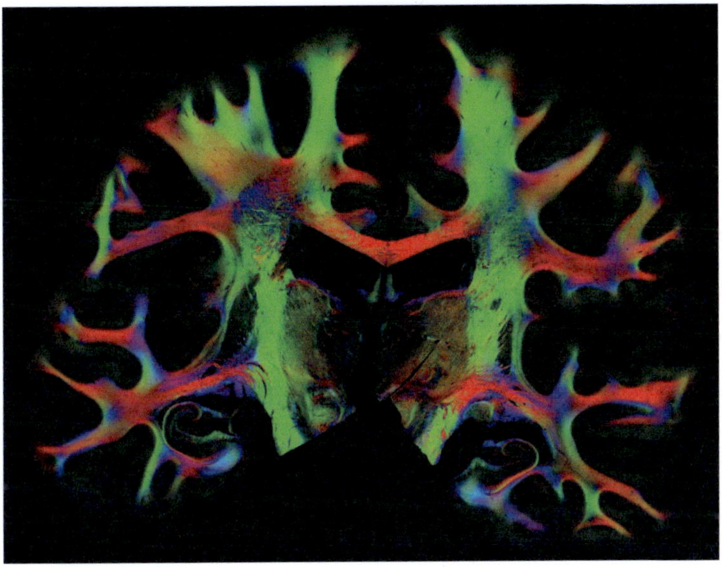

Figure 28: Fiber orientation map of a coronal human brain section generated by the 3D-PLI workflow.

runtime information of UNICORE compute jobs and workflows and stores it in searchable repositories. A first release is expected in the end of 2016.

4 Conclusion and Contributions

In conclusion, the DLCL Health contributed successfully implementing an efficient and fast analysis workflow for high-resolution 3D-PLI images, which is a prerequisite for large-scale human brain processing. By means of UNICORE, complex data analysis and high-performance computing was combined in an easy-to-use manner, thus, providing a versatile tool enabling workflow modifications (i.e., plugin of new software, optimization of parameter sets, or change of flow diagram) as well as reproducible routine data production. Finally, the UNICORE platform as a whole benefited from the complex 3D-PLI use case in that new features have been added to its workflow engine improving its suitability in the neuroscience domain and the scientific environment in general [Hag14].

Acknowledgments

Most tools in the workflow are developed at Forschungszentrum Jülich in the Fibre Architecture group led by Markus Axer and in close collaboration with Oliver Bücker and the Simulation Laboratory Neuroscience (both JSC).

Quoted LSDMA Publications

[Hag14] B. Hagemeier et al. "A Workflow for Polarized Light Imaging Using UNICORE Workflow Services". In: UNICORE Summit. Poznan, Poland, 2014.

Other References

[11d] *Jülich Dedicated GPU Environment*. 2011. URL: http://www.fz-juelich.de/ias/jsc/EN/Expertise/Supercomputers/JUDGE/JUDGE%5C_node.html.

[15c] *Jülich Research on Exascale Cluster Architectures*. 2015. URL: http://www.fz-juelich.de/ias/jsc/EN/Expertise/Supercomputers/JURECA/JURECA%5C_node.html.

[Axe11] M. et al. Axer. "A novel approach to the human connectome: Ultra-high resolution mapping of fiber tract in the brain". In: *NeuroImage* (2011): *vol.54, no. 2*, pp. 1091–1101.

[Dem10] Bastian Demuth et al. "The UNICORE Rich Client: Facilitating the Automated Execution of Scientific Workflows". In: *e-Science (e-Science), 2010 IEEE Sixth International Conference on*. IEEE. 2010, pp. 238–245.

[Str10] A. Streit et al. "UNICORE 6 - Recent and Future Advancements". In: *Annals of Telecommunications-annales des Télécommunications* (2010), pp. 757–762.

Federated Authentication and Authorization Infrastructure for LSDMA

Marcus Hardt[a], Arsen Hayrapetyan[a], Patrick Fuhrmann[b], Paul Millar[b]

[a] Karlsruhe Institute of Technology (KIT), Karlsruhe
[b] Deutsches Elektronen-Synchrotron DESY, Hamburg

Abstract Within the Federated Identity Management work package of DSIT we analysed the requirements of our users regarding federated authentication and authorization components. Based on these components an integrative architecture was developed. Several pilots have been implemented to demonstrate the feasibility and general usefulness of the proposed framework. The LSDMA AAI includes bridging between SAML, OIDC and X.509 infrastructures as well as support for console access for traditionally web-oriented protocols like SAML and OIDC.

1 Introduction

Modern scientific collaborations are dynamic federations of research institutions. They usually span geographically distributed domains with different security policies and security infrastructures. Two key requirements which enable successful research in these collaborations are *secure data sharing* and *usability* of secure data management tools. These requirements are partly incompatible, and finding an effective technical solution effectively satisfying both of them is a non-trivial task. The challenge is further exacerbated by the sheer size of the scientific data available to collaborations nowadays.
The secure data sharing frameworks that existed at the time the LSDMA project started were rather focused on stringent security requirements. For

instance, the International Grid Trust Federation (IGTF) had established a world-wide X.509 based Public Key Infrastructure, across which it had ensured high-level security assurance by requiring its participants to adhere to minimum requirements of operational security policy. X.509 became the de facto standard mechanism for secure data sharing in the academic world. However, the effort to manage the X.509-based security credentials and the identity vetting procedures of IGTF were perceived by the users as requiring substantial efforts and solutions were being sought to improve the user-friendliness of the infrastructure.
Another major standard approach to secure data sharing in academia were the identity federations based on the Security Assertion Markup Language (SAML) standard technology. Harnessing the expressive power of SAML to support numerous security policies, these federations do not prescribe a common operational security policy to its member institutions. Instead, they rely on the rigorousness of security policies practiced by each member organisation. One crucial aspect of SAML implementations that made its adoption widespread isthe facility of its integration with organisations' identity management systems. Although the security assurance level in a SAML-based identity federation can vary, the usability of the infrastructure is usually perceived by the users to be higher than that of the X.509 standard based ones. Thus, the main challenge of the LSDMA project in designing its Authentication and Authorization Infrastructure (AAI) was to support secure data sharing for its participants by establishing bridges between the X.509 standard based infrastructures on one hand, and the SAML-based identity federations on the other hand.

2 Requirements

LSDMA is driven by requirements of the communities we support. Therefore, at the start of the project, we initiated a requirements analysis in order to evaluate which boundary conditions our developments had to meet. The key elements found were:

- *Non web-based access*: In general, the researchers ran data analysis tasks either via web-based portals or command line tools. In X.509 based infrastructures the access to the data had been traditionally non web based, and majority of the data access tools had been implemented as command-line utilities. In SAML-based infrastructures the web-based authentication and authorization was prevalent, not least due to the fact that the Enhanced Client or Proxy (ECP) profile of SAML which standardizes authentication and authorization for non web based access, was not widely adopted and deployed in existing federations.
- *Federated access*: Communities with existing AAI needed to collaborate with other communities. Modifying existing security policies and mechanisms was prohibitive. Therefore, a technical solution that would enable seemless integration into an identity federation was requied. The communities that had not had established AAI also requested the possibity of federated access.
- *User friendly access*: The tools for secure data sharing and credential management had to be easy to learn and use.
- *Datacenter friendliness*: The AAI components had to be chosen so that they would not require substantial effort for integration into the community's data handling infrastructure. The existing AAI components had to be used wherever possible, minimizing the number of extra components to be deployed.
- *Support for SAML authentication and authorization*: Communities that already had SAML-based authentication available to their users wanted to have support for their services in the LSDMA AAI. Those communities that did not have a SAML-based solution in place still wanted to keep the option of using a SAML-based AAI open.

- *Support for X.509 based data access*: Many communities had had protected the access to their resources with X.509 based security infrastructures, so support for X.509 based data access was a natural requirement for them. Some of the communities wanted to benefit from the high security level associated with X.509 based authentication and authorization. Yet other communities wanted to be able to access X.509-protected data of their collaborators.

One important requirement for data sharing was the support for multiple authentication methods. In a relevant scenario an entity (user or an automated process) would authenticate with one type of security credentials and write data to the data storage and later would authenticate with another type of the credential to read the data. To enable this kind of data sharing, both types of entity security credentials (entity's different identities) would need to be mapped to the same entity or account for proper authorization by the underlying storage system, normally the (UID, GIDs) pair for a filesystem.

3 Scope

In order to get the idea of the size of the scope of secure data sharing and federated authorization one should consider it not only from the technical point of view, but also from the standpoint of different *trust models* employed by scientific collaboration AAIs. One largely employed trust model implies that an identity provider (IdP) is operated at each university or institute in the federation. Each of the IdPs has its own policy subject to the privacy laws of the host country and organisation. Another prominent trust model is based on the federation of certification authorities (CAs) with common minimum requirements for the operational policy.
Any data sharing solution proposed by LSDMA had to address not only the technical issues, but also be flexible enough to accomodate multiple combinations of security policies in the federation. To facilitate its adoption, the

proposed secure data sharing solution had to reuse the existing authentication and authorization frameworks (SAML IdPs, X.509 Certification Authorities, etc.) as much as possible.
At the beginning of the LSDMA project several large AAIs already existed. They involved stakeholders such as national research and education networks (NRENs), the pan-european collaboration on e-infrastructure and services for research and education, Geant, which develops the eduGAIN interfederation identity service, the European Grid Infrastructure (EGI) which had been using the IGTF PKI to provide authentication and authorization services to its users, as well as CERN (WLCG grid) and ESA (G-POD grid). These large AAIs were generally based on either SAML-based identity federations or X.509 based PKIs and were also actively seeking to integrate support for both.
Since the issues being addressed by LSDMA AAI are general and have been in the centre of attention of many different collaborations, it was natural choice for LSDMA to cooperate with projects seeking solutions in the same problem domain. These included INDIGO [16p], AARC [16c] and EU-DAT [15b].

4 Pilots

In order to demonstrate the viability of our proposal for LSDMA AAI we conducted several pilots. Each pilot focused on a particular aspect of the proposed infrastructure.

4.1 Credential Translation between SAML and X.509 based Systems for Cross-protocol Access

Using the test and production IdPs at KIT as SAML credential providers we wanted to employ an online certificate authority (CA) service to generate X.509 certificates for the users. [Har14] For this we employed the Short Lived Credential Service (SLCS) provided by the German National Research

Network provider (DFN). DFN SLCS is offered by the DFN-SLCS certification authority which is accredited by the Internation Grid Trust Federation (IGTF). DFN SLCS acts as a SAML service provider that accepts certificate signing requests from users that authenticated with their home-IdP. The users' home-IdPs are typically members of the German national identity federations (DFN AAI and DFN-AAI-Test). The production and test IdPs at KIT are members in these federations.

For each of the translations we have considered two options of authentication: web-based and non web-based. The web-based option is the default mode of operation of DFN SLCS CA. This option required the execution of a local java webstart application which handles certificate creation at the client computer. The non web-based option, however, was not supported. To communicate with the DFN SLCS CA in non web mode, we used the Enhanced Client or Proxy (ECP) profile of the SAML specification and asked DFN SLCS CA to change the configuration to support SAML ECP requests. However, as the investigation revealed, a number of further changes had to be applied to the DFN SLCS CA configuration. In addition, the codebase was deemed inappropriate for the production service. We have concluded that this approach requires substantial deployment effort and did not satisfy the deployability reqirement of the AAI.

The additional policy-related effort and the required development work led to finalising this pilot in its architectural phase. The policy issues have been adressed within the REFEDS and AARC projects and resulted in the R&E (Research and Education) entity category to allow attribute release and in the SIRTFY/SNCTFY initiatives for operational security.

As an alternative, we have proposed an approach based on in situ credential translation by a token tranlsation service (TTS) operating in the same domain as the service to which the access is made with X.509 credentials. The TTS obtains the identity and group information of the authenticating user from SP/IdP Proxy and forges accordingly an appropriate access token for the tar-

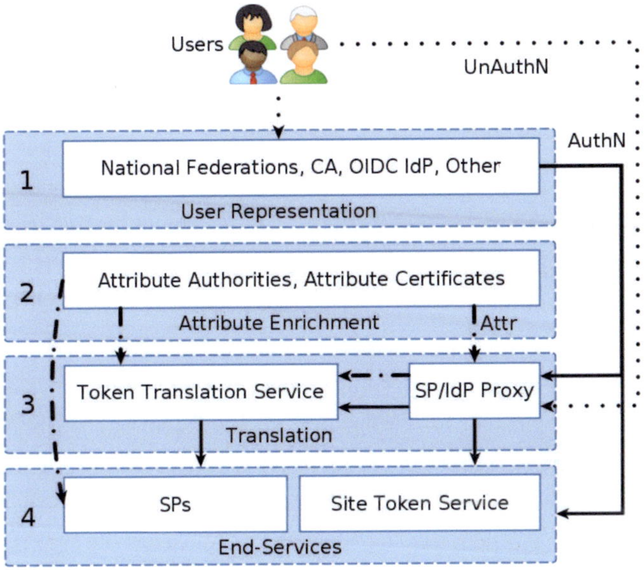

Figure 29: Credential translation-based authentication and authorization in the context of e-infrastructures.

get system, see Figure 29. This option is currently being implemented for gridftp test instance at KIT.

4.2 Federated Authentication of PAM-based Services

SAML, a widely used authentication solution for federated authentication, is mainly targeted towards authentication in web contexts. This is because the protocol depends on HTTP(S) redirects and allows for custom login procedures at the home-IdPs. However, many traditional Unix services such as ssh, imap, ftp, login or even sql databases facilitate the pluggable authentication modules (PAM) for authentication of users.

In this prototype we wanted to research the multi-protocol authentication for a non web-based services such as those using PAM for authentication. As an example we have made use of the well known service secure shell, *ssh*,

to see if a general PAM-based approach could support various authentication protocols like SAML, OpenID-Connect (OIDC, used by various public or social IdPs such as Google and Facebook).
Using PAM-LDAP an external LDAP server can be used for authorization. In such a scenario, the *ssh* server receives username and password from the user and pass them to an LDAP server to make the authorization decision. LDAP Facade, a solution, developed at KIT for federating HPC resources, leverages this possibility by using the Apache Directory LDAP Server, which (like many other LDAP servers) allows the registration of an interceptor for various functionalities. We have implemented a custom authentication interceptor ECPAuthenticator for PAM-based authentication.
LDAP Facade provides a component which implements the SAML authorization logic which is called by the interceptor. Two options are available. The user can provide *ssh* with their login name and password for their home-IdP. LDAP Facade will then use this password at their home-IdP for verification as per SAML ECP profile workflow.
Alternatively, the user can perform the login to an intermediate SP (using their home-IdP) upfront which generates a short lived token for the user. The users supplies the *ssh* server with this token (via the password field). Once arrived at the authentication intercepter, the token ca be verified with the intermediate issuing SP. The workflow is demonstrated on Figure 30.
For the first goal, we have succesfully demonstrated the possibility of enabling access to *ssh* using SAML, OIDC, and EUDAT's b2access (which in turn enables the user to authenticate via ORCID and other social IdPs such as Google). The SAML based PAM authentication (with ssh as an example) via the LDAP Facade solution has been succesfully deployed and was positively evaluated within the AARC project.

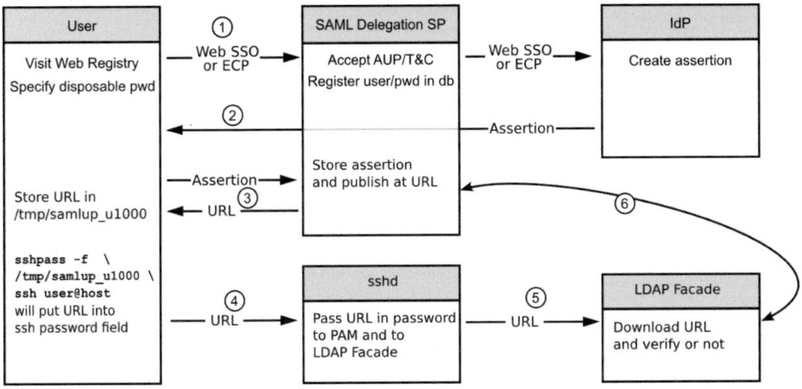

Figure 30: SAML-based ssh authentication with upfront assertion generation.

4.3 Federated Authentication of Kerberos based Services

Next to PAM, Kerberos is another popular authentication service for Unix. As a particular case we have developed a demonstrator, using LDAP Facade-enabled Kerberos authentication. For this we used an NFSv4 installation at KIT. In this scenario, the user authenticates to their home-IdP and gets their Kerberos key deployed into the LDAP directory server. A copy of this key needs to be delivered to the user who needs it to authenticate to the kerberos server to authenticate their NFS mounts. The detailed decription can be found in [Ber15].

4.4 Supporting multiple Federations and Guest IdPs

SAML allows to organise groups of IdPs and Service Providers (SPs) into federations. Today most federations exist on the national level. Each federation may have different policies that regulate the interplay of IdPs and SPs. Specifically, these policies may regulate handling of personally identifyable information or procedures for security incidents. EduGAIN is an interfederation that organises all missing links so that SPs in one country are capable of authenticating users from a different one.

Since SAML is typically found in the domain of research, there are cases, in which users do not have a home-IdP. In these cases some communities run their own IdPs, also called guest-IdP or homeless-IdP. Such IdPs are often not capable of providing the same information about users (attributes) than institutional IdPs.

With our pilot developments and deployments that used SAML authentication, we wanted to research the adaptability of these solutions in multiple different federations. Each federation comes with potentially different size, policy, and architecture. Even on the federation architecture level we found limitations for non web-based access.

We have succesfully demonstrated the integration of our *PAM* pilot (using again ssh) first with the bwIDM federation then also with the DFN-AAI, and the EduGAIN identity federations, but also with the Umbrella and the EUDAT B2ACCESS community driven IdPs.

Some federations found in EduGAIN have different architectures: for example, DFN-AAI is a full-mesh federation in which every participant organisation operates their own Identity Provider (IdP) and SAML assertions are directly released to SPs such as LDAP Facade. The Dutch SurfConnext is an example for a proxy federation. In that federation a single central IdP forwards queries to the users' home-IdP. This makes the use of ECP impossible for our case.

An important problem which arises in an identity federation (or interfederation) is that of the minimal set of IdP-asserted attributes requested by an SP to provide the service. Since every IdP has its own policy on which attributes to release, this number should be kept to a bare minimum, even zero, and user should be given a possibility to provide self-asserted required attributes. The SP might decide to provide limited service to a user who presents more self-asserted required attributes in comparison to the user who presents more IdP-asserted required attributes. However, the failure by the IdP to release required attributes should not result in the service being denied to the user.

We have added the support for this “zero-attribute-policy” to LDAP Facade and have succefully tested it with IdPs releasing only few or no attributes.

4.5 Identity Harmonization

As mentioned previously, the mapping of multiple user identities to a single entity (identity harmonization) is crucial to the secure data sharing. In the context of LDSMA we have discussed the problem and identified the architectural requirements of an identity harmonization service. The work formed a basis on which a full-fledged identity harmonization service (IDH) has been proposed and designed for INDIGO project. The conceptual and technical details of the service are presented in [Ert16].

5 Conclusion and Contributions

The key achievements of the LSDMA AAI have been presented. We have designed and implemented an Authentication and Authorization Infrastructure for secure data sharing, based on real-life security and usability requirements of LSDMA scientific communities, and demonstrated the feasability of our approach through the pilots we ran on the infrastructure. Following is the list of our key contributions:

- *Collection and analysis of security and usability requirements of LSDMA communities.* We captured and analysed the AAI requirements of participating communities. Thereby, we adressed various communities with different experiences in using an AAI. Together, we systematically specified their security, usability and deployability requirements. The requirements were categorised, analysed and turned into user-stories to follow an agile approach in addressing the users needs.
- *AAI design based on the identified requirements.* The general AAI prototype architecture was developed based on the requirements identified in the previous step. The architecture was designed with an inclusive

approach in mind, to span the whole range of AAI requirements of the participating communities. Corresponding technical means for the implementation of the prototype were sought.

- *Technical solution to secure data sharing problem.* The developed AAI prototype it included support for different authentication methods and protocols, and enabled cross-protocol access by proposing a functional mapping of the authenticated identities to the underlying storage system (UID, GIDs) pair. It also accounted for the management of this mapping, including the updates and unlinking of the identities. A proper identity mapping is critical for secure data sharing across storage systems with heterogeneous authentication and authorization mechanisms.

- *Evaluation of the developed prototype.* The general architecture was piloted with several different use cases that demonstrated the feasibility of the approach taken.

- *Cooperation with EU projects INDIGO-DataCloud and AARC.* Various components of the general architecture have been proposed for AAIs of EU projects seeking solutions to secure data sharing problem. The usefulness of the architecture was recognized by the INDIGO project, in which the proposed architecture forms one corner stone of the project's AAI. Furthermore, the LSDMA AAI was recognized by the AARC project, which includes it in the list of analysed architectures in its blueprint architectures document [16a].

Quoted LSDMA Publications

[Ber15] A. A. Bersenev et al. "An approach for integrating kerberized non web-based services with web-based identity federations". In: *Proceedings of the 10th International Conference on Software Paradigm Trends, ICSOFT 2015*. doi: 10.5220/0005509901440150. SCITEPRESS. 2015, pp. 144–150.

[Ert16] Benjamin Ertl et al. "Identity Harmonization for Federated HPC, Grid and Cloud Services". In: *Proceedings of the 2016 International Conference on High Performance Computing and Simulation*. IEEE. 2016, pp. 621–627.

[Har14] M. Hardt et al. "Combining the X.509 and the SAML Federated Identity Management Systems". In: *Proceedings of the 2nd International Conference, SNDS 2014 on Recent Trends in Computer Networks and Distributed Systems Security*. doi:10.1007/978-3-642-54525-2_36. Springer. 2014, pp. 404–415.

Other References

[15b] *EUDAT - A Pan-european Data Infrastructure*. 2015. URL: https://www.eudat.eu/.

[16a] *AARC Blueprint Architecture*. June 29, 2016. URL: https://goo.gl/imZFch.

[16c] *Authentication and Authorisation for Research and Collaboration*. June 29, 2016. URL: https://aarc-project.eu/.

[16p] *INDIGO - DataCloud*. June 29, 2016. URL: https://www.indigo-datacloud.eu/.

Federated Storage Infrastructure for LSDMA

Benjamin Erlt[b] Patrick Fuhrmann[a] Marcus Hardt[b] Paul Millar[a] Tigran Mkrtchan[a] Karsten Schwank[a] Marina Sahakya[a] Bas Wegh[b]

[a] Deutsches Elektronen-Synchrotron DESY, Hamburg
[b] Karlsruhe Institute of Technology (KIT), Karlsruhe

Abstract The high level objective of the data area within LSMDA is the provisioning of tools for conveniently storing, federating, accessing and sharing huge quantities of data. The resulting toolbox is mainly targeting scientific communities, not willing or not being able to develop their entire data management framework themselves. The selection of services and products within that toolbox is based by their usage of Open Standards and their availability on the Open Source market. Even more importantly, significant focus has been put on the evaluation of the potential self-sustainability of components, due to an active user community or due to the commitment of the product teams to further maintain their products, independently of LSMDA funding. Besides integrating well established and sustainable data management components, LSDMA evaluated gaps in existing data management procedures and, in response, either established working groups in international scientific organizations, like RDA or joined existing taskforces in industry, like SNIA, on those topics. As those activities naturally require agreements on the European and possibly international level, LSDMA partners successfully joined European projects, like the INDIGO-DataCloud or AARC, engaging a larger group of communities.

1 The Big Picture

The fundamental issue in federating storage in the highly heterogeneous environment, like the German University and Research Center infrastructure, is the diversity of access mechanisms to data endpoints, both in terms of authentication and identity management and on data transport and control protocols.

Although the AAI aspect is mostly covered by the corresponding work packages, it has significant implications on the storage system as well. In order to keep LSDMA solutions generic and essentially non-intrusive, a focus on open standards in all areas where inter-site communication is required, is essential. As for authentication *SAML* [01b] and " *OpenID Connect* " [14c] are the mechanisms of choice, in data access http/WebDAV for web applications, NFS4.1/pNFS [10b] for low latency high throughput POSIX access and GridFTP [05] for wide area bulk data transfer are covering most of the LSDMA use cases and are provided by industry as well as by community specific storage technologies in Germany.

A third area, besides authentication and standardized data access, is the ability to control the quality of storage used for the various steps within the scientific data lifecycle. While different technology and cloud storage vendors provide data storage with selectable quality attributes, like fast access or long term archiving, there is no standard mechanism to specify the requested type of storage in a programmatic way. Based on the experience we gathered while developing and deploying the Storage Resource Manager protocol (SRM) [09] in High Energy Physics, we started an initiative in collaboration with the Storage Network Industry Association, SNIA [97a] and the Research Data Alliance, RDA [13d] to extend the existing Cloud Data Management Interface, CDMI [CDM13] to allow supporting our ideas on Quality of Service in storage . Due to the involvement of LSDMA partners in European projects, this initiative became part the *Description of Work* of the INDIGO-DataCloud [16p] H2020 project and is now being work-on in laboratories spread over Europe.

To support a larger area of the typical scientific data life cycle, scientists need to be given the ability to share their data with individuals, groups or the public within their own Research Centers or with remote institutions, without necessarily coping data. In that context, we picked a popular Open Source product, ownCloud [11f] and incorporated it with dCache, the storage backend component of the LSDMA toolbox, giving scientists *Sync'n Share* capabilities

for their scientific data though ownCloud, and at the same time providing the necessary data access and storage quality control standards, required for the scientific workflow, through dCache.
Finally, to complete the LSDMA data management toolkit, we selected two well established tools from HEP: FTS [16n], a technology to transfer bulk data between storage systems controllable via a Web Interface or a RESTful API, and DynaFed [16k], a system to federate distributed storage into a single virtual namespace.

2 The Quality of Service in Storage Initiative

If there was affordable storage media, providing SSD-like low latency, high speed access and at the same time offering long term archiving features, the QoS initiative would become superfluous. However, there is not yet.
Consequently, the goal is to enable scientists or frameworks, through a standard API, to specify the storage quality needed for particular applications or dataset collections. Moreover, it is desirable to allow media transition following the scientific life cycle of the data.
Partners in LSDMA and INDIGO-DataCloud are working on a first prototype of this approach by extending an already existing cloud standard, the Cloud Data Management Interface, CDMI. The vocabulary of the interface is agreed on within a RDA working group and the technical implementation of the CDMI extension is specified together with SNIA, the maintainer of the CDMI protocol.
In order not to limit the development to only a limited number of back-ends, a CDMI framework protocol engine has been derived from the CDMI reference implementation, and is deployed by LSDMA. The different back-ends are managed though a plug-in system translating the CDMI request to back-end specific mechanisms, which are API's and scripts in case of products, like GPFS [98], HPSS [92] and CEPH [16g] or through a RESTful interface for dCache (see Figure 31). To track the development progress and to spot

regressions, an evaluation infrastructure has been setup, frequently querying the different CDMI endpoints in Germany and other European countries for supported CDMI operations. As a prove of concept and to gain initial experience, only disk and tape is initially supported.

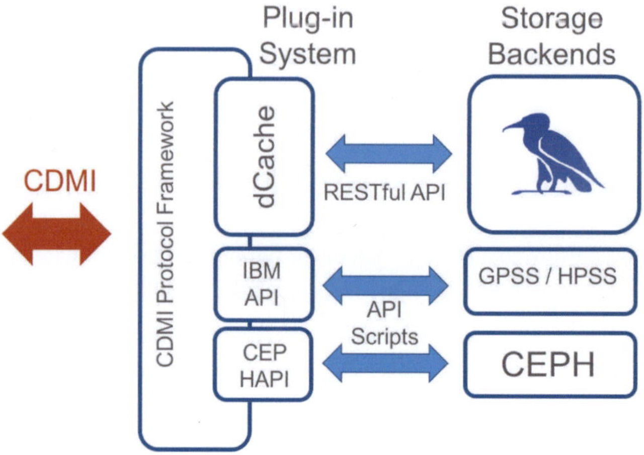

Figure 31: The LSMDA Quality of Service in storage.

3 Data Distribution and Federation

The FTS file transfer service can be combined with the DynaFed federation technology to satisfy a typical scientific use case. Data is produced at a sensor, in this case an radio antenna and the data is stored at a location close to the source. The experiment framework is using the RESTful interface of FTS to orchestrate the transfer of the raw data to a backup and an analysis site. At the analysis site, the data is processed and derived data is produced and subsequently transferred to the backup site. The content of all three repositories is federated using DynaFed, producing a virtual overlay file system from all three storage location. Data access requests through the federation system, using http/WeDAV, redirects the client to the geographically closest location for downloading the requested data. Using a more sophisticated mechanism, a client may decide to query for all possible location of the data and may decide to download datasets from all available locations at the same time to improve throughput. FTS and DynaFed are products developed and supported by the CERN storage group, and are in heavy use by the WLCG community.

4 dCache in LSDMA

dCache is the storage component of the LSDMA toolbox. The Open Source technology is developed, deployed and supported by dCache.org, a collaboration of DESY, FERMIlab [67] and the Nordic Data Grid Facility, NDGF [01a]. dCache is operating in production environments for more than a decade at about 70 sites around the world. Although initially designed to support HEP needs, it evolved to a general purpose storage system due to LSDMA funding, focusing on the support of standard data and authentication protocols. Most prominent contributions by LSDMA are the significant performance improvement of the NFS4.1/pNFS protocol, the integration of modern authentication mechanisms like SAML and OpenID Connect , the ability to aggregate small files on transition to tape improving tape performance, the integration of the ownCloud Sync'n Share features and the implementation of the QoS in storage features, discussed above.

4.1 dCache QoS in Storage Support

The support of the LSDMA storage QoS in dCache is achieved through a RESTful gateway. The CDMI protocol engine, described above, drives a dCache specific plug-in, converting the CDMI protocol to the dCache internal RESTful API. Besides supporting a variety of media types, the dCache resiliency module orchestrates service qualities whenever a predefined number of copies of the same file on different independent storage units is requested.

4.2 dCache Small File Support for Custodial Media Migration

dCache historically supports transitions of data between different media types, including custodial storage. The issue with those tape based systems is that due the non-sharable characteristic of tapes, tape drives and tape robot components, the offset time before the actual data transfer can begin is significantly larger than for random access devices. Depending on the load of the various components, those time offsets can be in the order of minutes to hours. Consequently, tape efficiency increases with the size of datasets moved from and to tapes. This is increasingly becoming an issue as with the evolvement of dCache from a big data store towards a general purpose system, the average file size is decreasing.
dCache is coping with the problem by introducing a file aggregation module. Data, before migrated to tape, is aggregated to large datasets which can be conveniently handled by the connected tape technology. The aggregation activity is transparent for the user. On recall of a single file, out of the aggregated collection, the big file is restored from tape and subsequently split into the original "small" files. Those small files are cached on disk as if restored from tape individually.

4.3 ownCloud and dCache

With the introduction of cloud type services and the inflation of mobile devices, the mechanisms of distributing and sharing data significantly changed over the previous ten years. User expect parts of their central data repositories to be automatically synchronized with their mobile devices in both directions. Similarly, when sharing a file with another individual or a group, the file is supposed to appear in a special directory of the target group in contrast to changing ACLs and pointing the target audience to the file in one's private file space. As in the scientific world, this feature is required on top of the traditional data management functionalities, the dCache team decided to in-

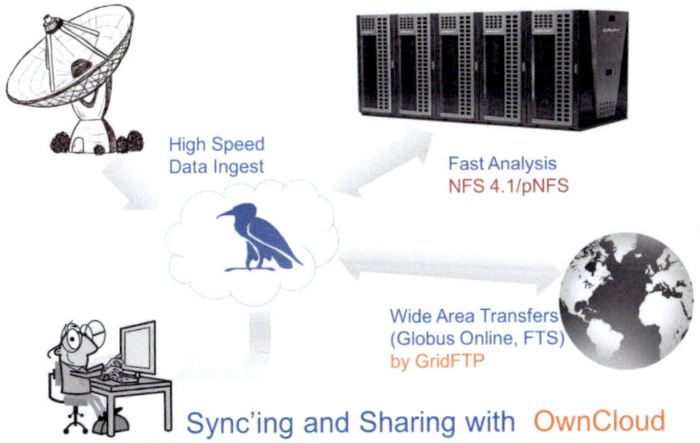

Figure 32: The dCache ownCloud hybrid System.

tegrate a popular Sync'n Share Open Source software with dCache to get the best of two worlds. The hybrid Is in production at DESY and SURFsara and is used by an increasing number of communities as their central Sync'n Share repository. Figure 32 sketches the various use cases of the hybrid system.

5 Executive Summary

During the elapsed funding period of LSDMA, the data area, responsible for storing, federating, accessing and sharing huge quantities of data, was able to collect and integrate a toolbox allowing scientific communities to build their own framework without reinventing and redeveloping basic storage tools. The selected components are Open Source and support Open Standards, guaranteeing non-intrusive integration into existing big data infrastructures. Sustainability is taking into account by only selecting technologies with a minimum successful history and an agile user community. In the area of QoS in storage, LSDMA is driving the bleeding edge of the state of the art, by collaborating with European projects, like INDIGO-DataCloud, global organizations, like the Research Data Alliance, RSA and industry bodies, like the Storage Networking Industry Association, SNIA.

Other References

[01a] *NDGF, Nordic Data Grid Facility*. 2001. URL: http://www.ndgf.org.

[01b] *SAML*. 2001. URL: https://de.wikipedia.org/wiki/Security%5C_Assertion%5C_Markup%5C_Language.

[05] *GridFTP V2*. 2005. URL: https://www.ogf.org/documents/GFD.47.pdf.

[09] *SRM, The Storage Resource Manager Interface*. 2009. URL: https://sdm.lbl.gov/srm/.

[10b] *NFS 4.1 / pNFS*. 2010. URL: http://www.pnfs.org/.

[11f] *ownCloud*. 2011. URL: https://owncloud.org/.

[13d] *RDA, Research Data Allicance*. 2013. URL: https://www.rd-alliance.org/node.

[14c] *OpenID Connect*. 2014. URL: http://openid.net/connect/.

[16g] *CEPH*. 2016. URL: http://www.Ceph.com.

[16k] *DynaFed*. 2016. URL: http://information-technology.web.cern.ch/.

[16n] *FTS, File Transfer Service*. 2016. URL: http://information-technology.web.cern.ch/services/file-transfer.

[16p] *INDIGO - DataCloud*. June 29, 2016. URL: https://www.indigo-datacloud.eu/.

[67] *FERMIlab*. 1967. URL: http://www.fnal.gov.

[92] *HPSS, Hierarchical Parallel Storage System*. 1992. URL: http://www.hpss-collaboration.org/.

[97a] *SNIA, Storage Networking Industry Association*. 1997. URL: http://www.snia.org/.

[98] *GPFS, General Parallel File System*. 1998. URL: https://en.wikipedia.org/wiki/IBM%5C_General%5C_Parallel%5C_File%5C_System.

[CDM13] CDMI. *Cloud Data Management Interface (CDMI)*. Sept. 2013. URL: http://www.snia.org/tech%5C_activities/standards/curr%5C_standards/cdmi/.

Advances in Metadata Research by LSDMA

Richard Grunzke[a], Ralph Müller-Pfefferkorn[a], Bernd Schuller[b], Stephan Kindermann[c], Carsten Ehbrecht[c], Thomas Jejkal[d], Volker Hartmann[d], Ajinkya Prabhune[d], Rainer Stotzka[d]

[a] Technische Universität Dresden, Dresden
[b] Forschungszentrum Jülich, Jülich Supercomputing Centre, Jülich
[c] German Climate Computing Center (DKRZ), Hamburg
[d] Karlsruhe Institute of Technology (KIT), Karlsruhe

Abstract Managing metadata in a generic way is of essential importance in scientific data life cycles. It needs to be both efficient and seamless. Such a concept was designed and implemented within the MoSGrid Science Gateway. The utilized UNICORE metadata management is a generic service within the UNICORE HPC middleware. The concept resulted in the DFG project MASi that utilizes the repository framework KIT Data Manager to built up a generic metadata-driven research data management service. Furthermore, metadata developments in general and provenance support specifically for the generic web processing framework birdhouse for the earth sciences are introduced. We highlight the valuable contributions in generic and specific metadata management that were designed and implemented.

1 Generic Metadata Handling in Scientific Data Life Cycles

A concept for generically handling metadata for scientific metadata was designed, implemented and evaluated in a doctoral dissertation [Gru16a] within LSDMA on which this section is based. Utilizing metadata, the approach facilitates thc move towards the next generation of data management within scientific data life cycles (see Figure 33). Metadata management in general enables the organization of large amounts of data with file number in the millions. Instead of only being accessible via names in directory structures, files can be accessed by their content. Despite the inherent complexity of data contents, the data can be seamlessly accessed via simple search queries and directly further utilized. The high complexity and large magnitude of scientific data life cycles is motivated. This subsequently necessitates sophisticated and integrated technologies for their management [Gru15]. Based on such technologies, scientists are enabled to advance their respective state-of-the-art with the combined support of High Performance Computing and Big Data resources. A multitude of open challenges in this broad data life cycle context was identified. Within the challenges, a major one is that of completely missing or only narrowly applicable metadata management approaches.

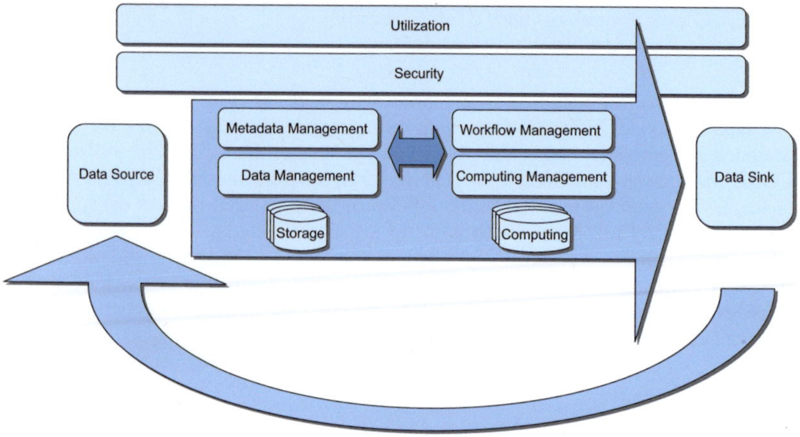

Figure 33: "Principal data life cycle management component categories are depicted with data source, storage and computing pillars, data sink, and layers for security and utilization. Each category consists of a multitude of technologies to manage the inherent complexity." [Gru16a]

The metadata concept [Gru14e; Gru14d] first specifies multiple advantageous characteristics. The abstraction of various technologies is heavily utilized to handle the high data life cycle complexity (see Figure 34). Data and metadata formats are integrated in a generic way to enable advanced search capabili-

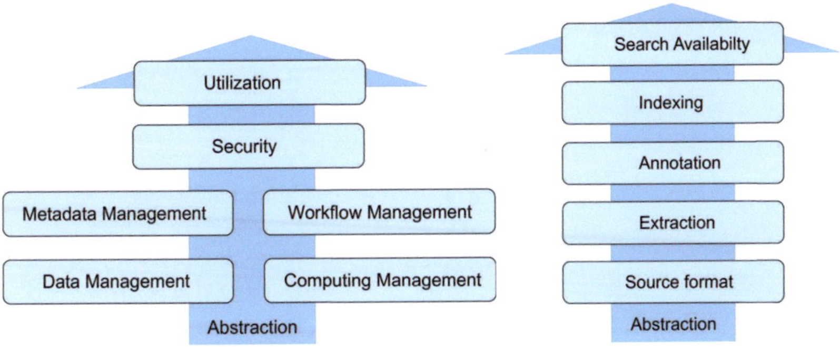

Figure 34: The left side displays the "abstraction from metadata systems and various other levels of abstraction from data and computing management up to the security and utilization layers are shown." [Gru16a] The right side "depicts various abstraction layers reaching from data and metadata formats at the base to search availability in ascending layers." [Gru16a]

ties for the efficient usage by scientists. The concept is inherently scalable within HPC-enabled and Big Data life cycles. Full automation is provided for extraction, annotation, and indexing of metadata. Second, the concept provides a design guide that facilitates an understanding of overall data life cycle design aspects [Ges15b] as well as aspects specific to metadata management. The guide includes recommendations of proven technologies. The overall concept is generic in the sense that it enables metadata management to be more quickly and efficiently integrated in concrete data life cycles. The direct usage of metadata search results is enabled while underlying Big Data and High Performance Computing resources are seamlessly integrated. Users gain from new capabilities while the added necessary complexity is hidden.
The concept was implemented within the MoSGrid data life cycle [Krü14b; Gru12; Ges14] MoSGrid enables highly complex molecular simulations within the three major computational chemistry domains. The MoSGrid implementation is HPC and workflow enabled, offers advanced data management capabilities with a sophisticated single sign-on architecture throughout all layers [Ges12b]. The implementation based on the generic approach was seamlessly integrated with this complex data life cycle. The meta-

data extraction, annotation, and indexing are performed in a fully automatic way [Gru14d]. A search interface enables finding data based on its content. Furthermore, search results can immediately be re-used as input within further workflows.

A thorough evaluation of the concept and its implementation was performed [Gru16b; Gru14d] with respect to adaptability, performance, sustainability, resilience, and efficiency of use. The existence of favorable properties was shown.

On a theoretical level, data life cycle management is advanced by facilitating higher level abstraction with metadata management. A practical impact is achieved by the implementation within the MoSGrid data life cycle and the uptake of the concept within the MASi research infrastructure.

2 UNICORE Metadata Service and Support in Clients

UNICORE [97c; Ben16] is a mature software for federating heterogeneous compute and data resources, comprising a full software stack from clients to various server components down to the components for accessing the actual compute or data resources. It offers several interfaces for realising data access and management, and is able to connect to a variety of backend systems that store the actual data and metadata.

The built-in support for metadata [NS10] consists of a metadata management interface which offers the typical *create*, *read*, *update* and *delete* methods. Metadata is modelled as simple key-value pairs, and there is no prescribed schema. This is done by the respective user community. Each UNICORE storage instance has its own metadata manager, which allows many application scenarios, depending on whether data is shared between users, or whether the data of different users should be separated. The metadata manager also allows to *search* for metadata using an implementation dependent query string, and it also allows to trigger automated *metadata extraction*.

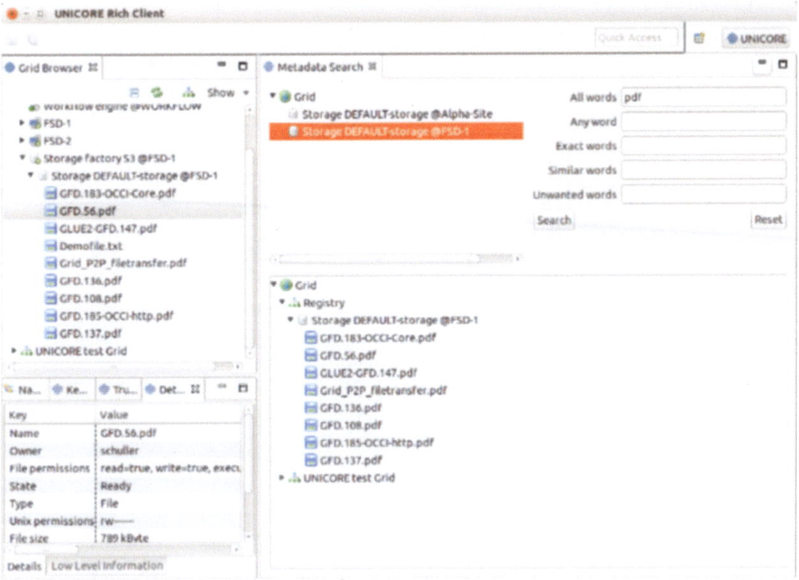

Figure 35: The metadata search view within the UNICORE Rich Client [Dem10].

The abstract metadata management interface comes with a default implementation, which stores the metadata on the same storage that also holds data. Metadata is simply written to hidden files. This approach has several advantages. No additional storage system (like a relational database) is required, and the metadata has the same access control restrictions as the actual data. Last but not least, administrative tasks like backup of the metadata do not require any additional effort.

Metadata is indexed using Apache Lucene [99], which also provides the search functionality. Metadata extraction is done using Apache Tika [07], which comes with a large number of built-in extraction filters for common file formats such as pdf or jpg. Apache Tika can be extended with custom extraction filters.

All the UNICORE Clients offer basic metadata support for managing metadata and for searching. It is also possible to access the metadata features using the new UNICORE REST API [SRB14].

3 An Integrated Research Data Repository Framework

If one takes a look at the scientific landscape there are hundreds of different metadata standards [Lan11] with many inside single communities. At Digital Curation Centre (DCC) a catalog exists with a rough overview of metadata standards categorized by communities [16j]. A widely accepted standard is Dublin Core (DC) [Wei98] defined in 1990. It defines a basic set of metadata which is most convenient for libraries. Nowadays it serves as common base for nearly all tools dealing with metadata. Although there exist a couple of extensions, this standard is of limited use for non-bibliographic data.

In LSDMA we are dealing with scientific data which cannot be sufficiently described with DC. To store and manage large scale research data we designed and implemented the KIT Data Manager (KIT DM)[Jej14; KIT16] which is a software framework for building up research data repositories. As part of KIT DM, a fixed set of administrative metadata was defined, which has to be part of each digital object stored in a repository instance. This set contains the minimum requirements for administrative metadata e.g. experimenter and date. There is also a possible transformation to DC in order to provide metadata in a standardized schema to external tools. Additionally, a metadata module extension is available for the KIT DM. This extension provides a framework which allows communities to extract community specific metadata during ingest. The extracted metadata is expected to be stored as XML. If a corresponding XML schema definition is available the metadata will be validated. If the extracted metadata is valid the ingest will be finalized and the metadata will be indexed, e.g. in Elasticsearch. If the validation fails, the ingest will be rejected. Even if no community specific metadata exists at

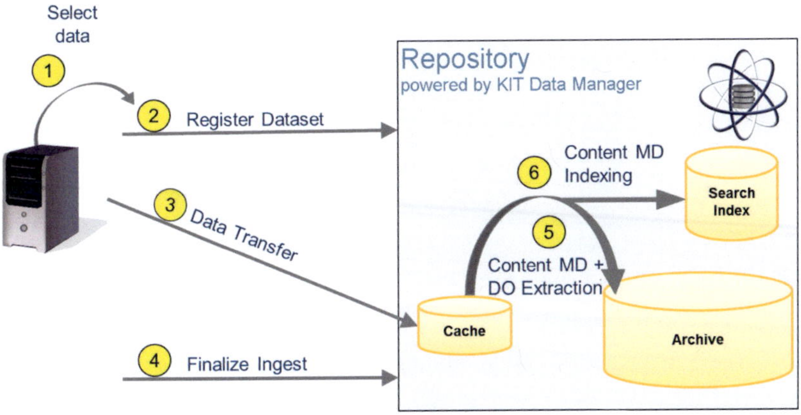

Figure 36: Data/Metadata workflow for ingest operations within KIT DM. After selecting the data (1) the administrative metadata for a new Digital Object is registered (2) and a new ingest is scheduled. As soon as the ingest is prepared, the data can be transferred by the user (3). Finally, the ingest is enabled for finalization (4). During finalization the data is moved to the archive (5), the Data Organization (DO) is obtained and content metadata might be extracted automatically. Finally, extracted content metadata is made searchable via a search index (6). [KIT16]

least the basic administrative metadata will be indexed. This ensures the retrievability of every digital object in the repository. Furthermore, extracted metadata can be harvested from within KIT DM via OAI-PMH [Van04]. OAI-PMH which is a standard protocol for harvesting metadata from data providers and allow value-added services for service providers.

To conclude, a generic metadata workflow within the KIT DM framework was defined. It supports automatic extraction of community specific metadata, validation, distribution, and retrievability. It is recommended for the communities in order to improve reusability of data. Inside the KIT DM, the metadata workflow is implemented and the feasability was proven by the integration for several scientific communities, e.g. arts and humanities, biology and energy research.

4 MASi - A Generic Metadata-driven Research Data Repository Service

MASi [Gru16c; MAS15] is a DFG project to build up a generic metadata-driven research data repository service. The successful proposal directly resulted from the close collaborations within LSDMA and specifically the metadata workpackage
MASi "provides a solution for the highly relevant challenge of managing large amounts of complex data. It extends the existing KIT Data Manager framework by a generic programming interface and by a generic graphical web interface. Advanced additional features includes the integration of provenance metadata and persistent identifiers. The MASi service aims at being easily adaptable for arbitrary communities with limited effort. The MASi research data management service is currently being built up to satisfy these complex and varying requirements in an efficient way." [Gru16c]
"As future work, we are evaluating the integration of MASi both with the UNICORE HPC middleware [Ben16] and science gateways such as MoSGrid [Krü14a]. We are also continuing to work within the RDA and contribute our own expertise in discussions to create RDA recommendations on how to best handle various aspects of research data management. We aim at implementing the resulting joint recommendations within MASi. This will contribute in the creation of MASi as a service that is efficient, future-proof and has a high user acceptance." [Gru16c]

5 Metadata Services supporting Earth Science Data Management and Processing

The LSDMA project significantly contributed to the development of a generic web processing framework supporting web based earth science data analysis. The framework is called "birdhouse" providing OGC Web Processing Service (WPS) standard [16s] conforming processing services and is managed as an open source project [16d]. The following section describes the metadata related aspects of the birdhouse framework as well as metadata related earth science application scenarios currently worked on. A generic overview of the birdhouse component system was given previously. Earth science data analysis workflows in general need metadata support to 1) manage input and output data (data discovery and data catalogs), 2) manage and interact with processing services (service discovery, service invocation, service catalogs) and 3) manage data provenance (information on actors + in/out data + processing activities).

The birdhouse framework includes components to access and catalog external data sources provided by thredds [Dom06] data servers [16w] including the Earth System Grid Federation (ESGF) data archive [11c]. Additionally, it includes a component to index and search large local data collections using Solr technology [16b; Gru14b]. Also dedicated metadata quality assurance processes are provided by birdhouse to e.g. check data with respect to conformance to the Netcdf-CF metadata convention as well as generic model intercomparison project metadata conventions (e.g. to support CMIP6). For managing and interacting with processing services, birdhouse uniformly exposes OGC WPS standard based interfaces. The OGC WPS interface descriptions can be registered in an OGC Web Catalog Service supporting standards based service discovery. Processing results can be published to the same Catalog Service. These components are illustrated in Figure 37.

Work on generic data provenance support services relies on stable data references as well as agreed upon APIs and formats to record provenance information. Establishing these is a long term goal which needs clear (short term) use cases on one hand side and longer term community discussions and agreements on the other side. With respect to stable data references, work concentrates on establishing a persistent identification infrastructure in the climate community supporting the next round of climate model intercomparison projects (CMIP6). This includes a strong involvement in related working groups of the Research Data Alliance (RDA) [15d].

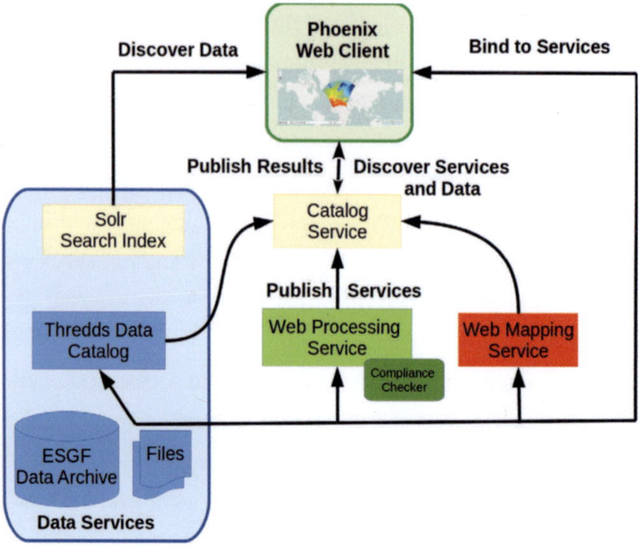

Figure 37: The birdhouse metadata architecture for earth sciences.

Discussions and prototype implementations with respect to provenance recording are proceeding in collaboration with the EUDAT project [Lec13; 15b] as well as the ENVRI+ project [15a] including cross-community discussions. Initial work follows the W3C PROV provenance metadata standard [MBC13; 15f].

6 Conclusion & Contributions

This chapter presents the metadata research that was done in the LSDMA project. A generic metadata management concept, published as a dissertation, was designed, implemented and evaluated. It utilizes the metadata service of the UNICORE HPC middleware. Its significant extensions within LSDMA were also described. Then, metadata functionality within the generic repository framework KIT Data Management was described, whose development was fundamentally enabled by LSDMA. Based on the close collaboration within the LSDMA metadata workpackage, the MASi DFG project proposal was conceived and subsequently granted. MASi is currently bulding a generic metadata-driven research data repository. Furthermore, substantial metadata advances were enabled in the earth sciences. To conclude, LSDMA enabled significant and far-reaching advances in various fields of metadata management.

Acknowledgements

Financial support by the German Research Foundation for the MASi project (NA711/9-1) is gratefully acknowledged.

Quoted LSDMA Publications

[16d] *Birdhouse Open Source Project, Collection of WPS related components to support Climate data processing*. 2016. URL: http://bird-house.github.io/.

[Ben16] Krzysztof Benedyczak et al. “UNICORE 7 - Middleware Services for Distributed and Federated Computing”. In: *International Conference on High Performance Computing Simulation (HPCS)*. 2016. DOI: 10.1109/HPCSim.2016.7568392. URL: http://dx.doi.org/10.1109/HPCSim.2016.7568392.

[Ges12b] Sandra Gesing et al. “A Single Sign-On Infrastructure for Science Gateways on a Use Case for Structural Bioinformatics”. In: *Journal of Grid Computing* 10.4 (2012), pp. 769–790. ISSN: 1570-7873. DOI: 10.1007/s10723-012-9247-y. URL: http://link.springer.com/article/10.1007%2Fs10723-012-9247-y.

[Ges14] Sandra Gesing et al. “Molecular Simulation Grid (MosGrid): A Science Gateway Tailored to the Molecular Simulation Community”. English. In: *Science Gateways for Distributed Computing Infrastructures*. Springer International Publishing, 2014, pp. 151–165. ISBN: 978-3-319-11267-1. DOI: 10.1007/978-3-319-11268-8_11.

[Ges15b] Sandra Gesing et al. “Science Gateways - Leveraging Modeling and Simulations in HPC Infrastructures via Increased Usability”. In: *International Conference on High Performance Computing Simulation (HPCS)*. July 2015, pp. 19–26. DOI: 10.1109/HPCSim.2015.7237017.

[Gru12] Richard Grunzke et al. “A Data Driven Science Gateway for Computational Workflows”. In: *UNICORE Summit 2012 Proceedings*. Vol. 15. IAS Series. 2012, pp. 35–49. ISBN: 978–3-89336-829-7. URL: http://hdl.handle.net/2128/4705.

[Gru14b] Richard Grunzke et al. “Device-driven metadata management solutions for scientific big data use cases”. In: *2014 22nd Euromicro International Conference on Parallel, Distributed, and Network-Based Processing*. IEEE. 2014, pp. 317–321.

[Gru14d] Richard Grunzke et al. “Standards-based Metadata Management for Molecular Simulations”. In: *Concurrency and Computation: Practice and Experience* 26(10) (2014), pp. 1744–1759. ISSN: 1532-0634. DOI: 10.1002/cpe.3116.

[Gru14e] Richard Grunzke et al. “Towards Generic Metadata Management in Distributed Science Gateway Infrastructures”. In: *IEEE/ACM CCGrid 2014 (14th International Symposium on Cluster, Cloud and Grid Computing)*. Chicago, IL, US, May 2014, pp. 566–570. DOI: 10.1109/CCGrid.2014.98.

[Gru15] Richard Grunzke et al. “Managing Complexity in Distributed Data Life Cycles Enhancing Scientific Discovery”. In: *IEEE 11th International Conference on e-Science*. Aug. 2015, pp. 371–380. DOI: 10.1109/eScience.2015.72.

[Gru16a] Richard Grunzke. “Generic Metadata Handling in Scientific Data Life Cycles”. PhD thesis. Doctoral Thesis, Technische Universität Dresden, Apr. 2016. URL: http://nbn-resolving.de/urn:nbn:de:bsz:14-qucosa-202070.

[Gru16b] Richard Grunzke et al. “Metadata Management in the MoSGrid Science Gateway - Evaluation and the Expansion of Quantum Chemistry Support”. In: *Journal of Grid Computing* (2016), pp. 1–13. ISSN: 1572-9184. DOI: 10.1007/s10723-016-9362-2. URL: http://dx.doi.org/10.1007/s10723-016-9362-2.

[Gru16c] Richard Grunzke et al. “Towards a Metadata-driven Multi-community Research Data Management Service”. In: *Proceedings of the 8th International Workshop on Science Gateways*

(IWSG 2016). Vol. 1871. CEUR-WS, 2016. URL: http://ceur-ws.org/Vol-1871/.

[Jej14] Thomas Jejkal et al. "KIT Data Manager: The Repository Architecture Enabling Cross-Disciplinary Research". In: *Large-Scale Data Management and Analysis (LSDMA) - Big Data in Science*. 2014, pp. 9–11. DOI: 10.5445/IR/1000043270.

[Krü14a] Jens Krüger et al. "Performance Studies on Distributed Virtual Screening". In: *BioMed Research International* (2014). DOI: 10.1155/2014/624024. URL: http://dx.doi.org/10.1155/2014/624024.

[Krü14b] Jens Krüger et al. "The MoSGrid Science Gateway - A Complete Solution for Molecular Simulations". In: *Journal of Chemical Theory and Computation* 10(6) (2014), pp. 2232–2245. DOI: 10.1021/ct500159h.

[SRB14] Bernd Schuller, Jedrzej Rybicki, and Krzysztof Benedyczak. "High-Performance Computing on the Web: Extending UNICORE with RESTful Interfaces". In: *Proceedings of the Sixth International Conference on Advances in Future Internet*. IARIA XPS Press, 2014, pp. 35–38. ISBN: 978-1-61208-377-3. URL: http://www.thinkmind.org/%5C-index.php?view=article%5C&articleid=afin%5C_2014%5C_2%5C_10%5C_40020.

Other References

[07] *Apache Tika Website*. 2007. URL: http://tika.apache.org/.

[11c] *Earth System Grid Federation*. 2011. URL: http://esgf.llnl.gov/.

[15a] *Envriplus - Integrated Solutions for Environmental Research*. 2015. URL: http://www.envriplus.eu/.

[15b] *EUDAT - A Pan-european Data Infrastructure*. 2015. URL: https://www.eudat.eu/.

[15d] *Research Data Alliance (RDA) - Research Data Sharing Without Barriers*. 2015. URL: https://rd-alliance.org/.

[15f] *W3C PROV Standard Overview*. 2015. URL: https://www.w3.org/TR/prov-overview/.

[16b] *Apache SOLR Open Source Enterprise Search Platform*. 2016. URL: http://lucene.apache.org/solr/.

[16j] *Digital Curation Centre - Metadata Standards*. 2016. URL: http://www.dcc.ac.uk/resources/metadata-standards.

[16s] *OGC Web Processing Service (WPS)*. 2016. URL: http://www.opengeospatial.org/standards/wps.

[16w] *Thredds Data Server*. 2016. URL: http://www.unidata.ucar.edu/software/thredds/current/tds/TDS.html.

[97c] *UNICORE Website*. Aug. 1997. URL: http://www.unicore.eu/.

[99] *Apache Lucene Website*. 1999. URL: http://lucene.apache.org/.

[Dem10] Bastian Demuth et al. "The UNICORE Rich Client: Facilitating the Automated Execution of Scientific Workflows". In: *e-Science (e-Science), 2010 IEEE Sixth International Conference on*. IEEE. 2010, pp. 238–245.

[Dom06] Ben Domenico et al. "Thematic Real-time Environmental Distributed Data Services (THREDDS): Incorporating Interactive Analysis Tools into NSDL". In: *Journal of Digital Information* 2.4 (2006).

[KIT16] KIT. *KIT Data Manager*. 2016. URL: http://datamanager.kit.edu.

[Lan11] Betty Landesman. "Seeing Standards: A Visualization of the Metadata Universe". In: *Technical Services Quarterly* 28.4 (2011), pp. 459–460. DOI: 10.1080/07317131.2011.598072.

[Lec13] Damien Lecarpentier et al. "EUDAT: A New Cross-Disciplinary Data Infrastructure for Science". In: *International Journal of Digital Curation* 8.1 (2013), pp. 279–287.

[MAS15] MASi. *Metadata Management for Applied Sciences*. 2015. URL: http://www.scientific-metadata.de/.

[MBC13] Paolo Missier, Khalid Belhajjame, and James Cheney. "The W3C PROV Family of Specifications for Modelling Provenance Metadata". In: *Proceedings of the 16th International Conference on Extending Database Technology*. ACM. 2013, pp. 773–776.

[NS10] Waquas Noor and Bernd Schuller. "MMF: A flexible framework for metadata management in UNICORE". In: *Proceedings of the 6th UNICORE Summit*. Vol. 5. 2010, pp. 51–60. URL: http://hdl.handle.net/2128/3812.

[Van04] Herbert Van de Sompel et al. "Resource Harvesting within the OAI-PMH Framework". In: *D-lib magazine* 10.12 (2004), pp. 1082–9873. URL: http://www.dlib.org/dlib/december04/vandesompel/12vandesompel.html.

[Wei98] Stuart Weibel et al. "Dublin Core Metadata for Resource Discovery". In: *Internet Engineering Task Force RFC* 2413 (1998), p. 222.

Long-term Access to Data, Communities, Developments and Infrastructure

Jos van Wezel[a], Felix Bach[a], Peter Krauss[a], Tobias Kurze[a], Jörg Meyer[a], Ralph Müller-Pfefferkorn[b], Jan Potthoff[a]

[a] Karlsruhe Institute of Technology (KIT), Karlsruhe
[b] Technische Universität Dresden, Dresden

Abstract Ubiquitous access to existing scientific data is one of the key enablers for rapid development and progress of empirical and without doubt, also non-empirical sciences. Management, in particular the long-term preservation of said data presents a major challenge.
Universities, scientific and cultural organisations, international collaborations and projects have a need to preserve and maintain (open) access to large volumes of digital data for several decennia. Existing systems supporting these requirements span from simple databases at libraries to complex multi-tier software environments developed within scientific communities and organisations. Generally, all communities are confronted with a high volume of data that must be handled efficiently and with increasing data volumes, also economically.
Development and integration of components to enable secure and reliable archival storage that make use of existing computer centres and infrastructures is an intrinsic goal in LSDMA and the project generally contributes to the improvement and development of support for long time scientific and non-scientific data preservation. Partners in LSDMA represent multiple scientific domains and are active in a range of international projects on data management. Collected requirements from partners are used as input to accompanying initiatives that cover a much wider scope as compared to community bound projects.
Although various national and international projects aiming to improve common practice of long term preservation have ended successfully, shortcomings still exist. E.g. even basic procedures such as data exchange between different systems are still very difficult to perform. The EU projects supporting infrastructure for research data focused this far on large-volume data sets and has promoted more organizational measures under the Research Data Alliance. Although the results are most important components of a comprehensive solution to the challenges, they cover often only se-

lected problems such as the preservation of small volume research data. The challenge identified by LSDMA is to connect island solutions with each other and thus to contribute to an integrated and easy to use system for data archiving and publication for researchers.

1 The Data Centre and its Role in Data Preservation

Data-centres have played a pivotal role in providing reliable data storage for universities and research facilities. Where libraries and archives traditionally governed the safe keeping of scientific paper based data, the data centre stored and preserved data from simulations, measurements, recordings, observations that result from the use of high performance computing and experiments involving advanced data taking instruments. Post publication data associated with simulations and experiments must remain accessible for the workgroup, the community and increasingly also for a knowledgeable audience and if properly presented, for a wider audience as in the case for open data . The technologies used for storing, retrieving and searching data require experts and specialisations ranging from software development to storage administration.

The huge increase in data volume that needs to be archived inevitably involves the data centre that must host cost efficient, established and trusted storage technologies but also provide the development of workflows and software that allow volume with the rising demand for archive storage. At the same time innovations, be it in house developed or market driven, must accommodate existing infrastructures, practices and applications. Today the physical data storage is usually a combination of magnetic disk and magnetic tape. Only large data centres, which includes commercial cloud providers can leverage the, compared to disk storage, higher reliability and economic benefits of tape.

In this article an archive equals the data center infrastructure which provides long-term storage for scientifically or historically important data that require no immediate access. The collection of archived objects that logically belong together or are managed by a single application instance is usually named a repository. To limit confusion and to remain in line with the name of the work package of LSDMA we use the more generic term 'archive' here.

2 Users of Archive Technology

2.1 Libraries

Goal of the state funded project bwDataDiss [16e] is to deliver services for Ph.D. students in the context of their thesis. Besides the doctoral dissertation Ph.D. students often produce research data that is necessary to understand and to reproduce findings of the thesis. The amount of research data may vary greatly and libraries usually lack the means to store the associated research data. bwDataDiss enables university libraries in the State of Baden-Württemberg to store and archive research data alongside the final dissertation.

The research data is preserved for at least ten years and accessible to other researches and the general public. The project promotes and is committed to Open Access [16t]. Two important characteristics of bwDataDiss are integrity assurance and embargos. Users expect integrity of data at all times. In particular, we calculate checksums every time the data is transferred from one system to another. This applies to transfers from the personal resp. office computer of the Ph.D. student as well. The system allows apply 'embargos' i.e. timespans during which no public access allowed. This is a requirement of libraries and a common practice to prevent access to publications for example pending patent application.

Project bwDataDiss plans to enable uploads to a size in the order of 10GiB and guarantees long term storage currently for at least ten years following the guideline for good scientific practice [16o]. For this it relies on storage services [PW14] of the bwDataArchiv long time archival service of the SCC data centre. The service must at least provide the following functionalities for bwDataDiss:

- Assurance of data integrity which is important for trusted reliability of the storage service. The archive, resp. the long term storage service must provide information about the integrity of the research data. The long term storage service should provide integrity information on a regular basis to ensure that the data is still correctly stored in the archive.
- Low access latency on frequently accessed small files. Even though a long term storage service cannot be considered as common storage that normally provides quick response times, bwDataDiss requires low latency access to certain content of the archived data. This is a requirement of the library, trying to provide a good user experience. If the storage service uses tapes instead of disks access time will be higher. bwDataDiss needs the ability to selectively store small files on disk and larger files on tape as this is more cost effective. Small files will be accessed more frequently compared to complete datasets stored in the deep archive.
- Write and delete protection for archived data. It must not be possible to accidently modify or delete data in the long term storage service.

The system developed in bwDataDiss is currently being tested and first data ingest is expected in early 2017.

2.2 Climate Research

In climate research it is very important to analyse historic data over long time periods in order to be able to model today's climate and to predict its evolution and long-time changes in the future. There are an increasing number of instruments and observational platforms on ground-based stations, air craft, balloons, and satellites. Observational data are very valuable not only for its scientific use and the effort that went into the acquisition, but also because a measurement at a given point in time cannot be repeated. That is why the demand on long-term archival for climate data usually is to preserve data forever, or at least much longer than the average preservation time for other scientific data.

Additionally is desirable to store processed climate data of research institutes on nearby facilities with a high bandwidth network connection. The Institute for Meteorology and Climate Research (IMK) at KIT, stores valuable data in the LSDF operated by the SCC data centre [Gar11b]. Although the LSDF provides online storage and the possibility to keep several replicas, IMK and SCC independently store databases guarded with checksums to ensure proper bit preservation. The copies provide for an additional level of data protection in case of problems with systems on either of the sites. While climate researchers can easily validate online data on disks by recalculating the corresponding checksums, the computing centre reluctantly lets scientists verify the data integrity of data on tape which can be handled more efficiently inside the data centre. An improved method for continuous check summing is presented at the end of this article.

2.3 HPC

In HPC data centers the amount of data stored is typically in the size of petabytes. Interestingly the growth of archival data is proportional to the amount of main memory of an HPC system. A survey from 2010 [Hic10] among data centres in the US estimates that for each Byte of main memory 35 Bytes of data is archived per year. The number of archived files is proportional to the number of cores in an HPC system, however no overall planning number could be estimated in the survey. At HLRS, the amount of archive data is growing continuously and holds approximately 6 PB data on

two copies making the actual amount of data on tape more than 12 PB. The estimate of the proportion 1:35 from the survey, is 1:20 at the HPC systems of HLRS, probably due to difference in the communities using the system.
After acceptance of a project proposal and access is granted, users start processing whereby data is produced and written to the filesystem. Although different workflows can be observed, commonly the data is post-processed, in which parts of the data is selected, aggregated, shrunken and sometimes even re-produced or visualized to gather knowledge on the problem. Subsequently the results are published and at the end of the project, data in the project workspace is no longer active and finally has to be removed to make room for further projects. However access to the data must remain possible for at least several years depending on the content and the requirements of the researchers.
A typical large project at HLRS generates some 10^5 files, each up to several Gigabytes in size. Main research areas are engineering, CFD, molecular dynamics and climate research with users originating from all over Europe. Given that each of these research areas show an almost exponential requirement for data processing with accompanying data growth, the results is increasing archive requirements. Since for processing, it is not necessary to keep this data on site, HLRS is working together with KIT for an offsite solution for large scale data archival. The archival requirements of HLRS start from 0.5 PB to some 2.5 PB long time storage in 2018.
sectionMulti-disciplenary archival

2.4 RADAR - Research Data Repository

Aim of this project [16u] is to establish and deploy a generic research data infrastructure in Germany. For this purpose, appropriate services for archiving and publication of research data are developed. The Steinbuch Centre for Computing (SCC), the project additionally involves the Leibniz Institute for Information Infrastructure (FIZ) at Karlsruhe, the Technical Information Li-

brary (TIB) in Hanover, the Leibniz Institute of Plant Biochemistry (IPB) and the Ludwig Maximilian University of Munich (LMU).
Due to the long term nature of the services, the independence between the archive management and the archive storage layer is an important operational benefit available in RADAR. Independency is ensured through clearly defined and standardized interfaces between both layers and these must provide, in addition to the pure data input and output, additional functionality such as checking the integrity, setting retention periods and especially the exchange of metadata that is specific to the management of the data inside the archive. Furthermore RADAR seeks to establish a contractual foundation for subscription based data storage which involves costs models and service level agreements between providers of data and of services.
Established as a national generic end-point repository, the RADAR particularly supports communities lacking their own community specific data archive and functions as catch all for other data worth archiving i.c. data related to scientific publishing. Its focus is therefore primarily on interdisciplinary institutions and the long tail of scientific data. Already included are (university) libraries and their associations as well as scientific publishers and cultural organizations.
Basically RADAR stores data, comprising scientific primary data and descriptive meta data, in data packages that may contain one or more files. Data packets are guarded against changes (bit preservation) and are stored for the duration the user requested. The repository is managed with the help of a web accessible user interface or a REST based application programmers interface (API). Currently RADAR uses the SCC data centre to store data.

2.5 OpARA - Open Access Repository and Archive

In the project OpARA the TU Bergakademie Freiberg and the TU Dresden build up a cross-disciplinary repository for research data for the state of Saxony. It allows scientists to describe their data with sufficient metadata, to

archive, and to publish them. In the spirit of Open Science the data can be made publicly available, but access can be restricted or embargoed e.g. if intellectual property rights are involved and the user chooses to postpone publication. Digital identifiers (DOI) are used to reference data sets.
The (meta)data is searchable, and a public interface based on OAI-PMH allows harvesting of metadata by external services. OpARA is integrated into the university environment by using the local identity management. Synchronization with other data services, like the university research information system holding research project information, is planned to improve data consistency. Researchers can ingest data either via a web browser or with a command line client. The latter is especially useful for large data sets or the recurrent ingest of similar data. OpARA is funded by the Saxon Ministry of the Arts and Sciences.

2.6 bwDIM

Crossing archive borders and enabling exchange of archived data between systems, communities and publishers is one of the main goals of the project 'data in motion' (bwDIM). Existing infrastructures and established repositories are locked-in solutions which contrast with the promises of long time archives. Since even the most reliable archive may and will be decommissioned, a means to exchange content with other repositories is required for redundancy reasons. The need for a data management exchange and an appropriate interface at the physical and application level will become more prominent as archives grow and exist longer. One of the goals in the project is the development of one of the first repository exchange interfaces.
Additionally the project develops a means to lightweight publication for large scale measurement data that is tightly connected to the RADAR repository. Because repositories such as RADAR must focus on curation and content preservation these are complex software packages that offer workflows, multiple data base interface, user and access control management and versioning

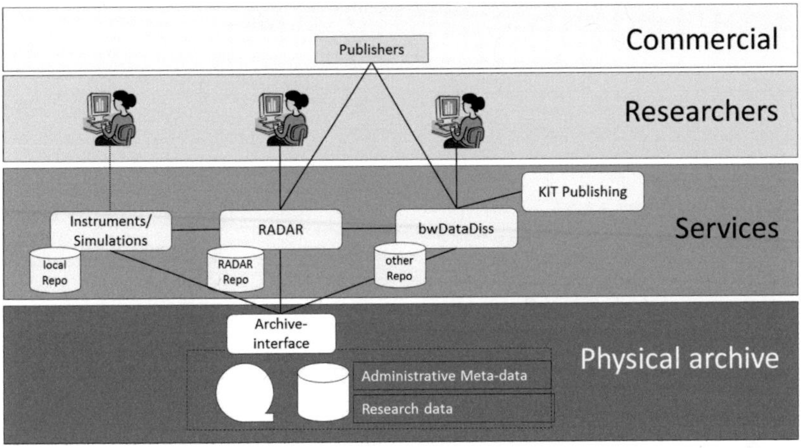

Figure 38: Data pathways and actors in bwDIM.

to name a few common features. Tying data processing, large scale, reliable storage and high speed networks from large data centers into the repository workflow of the RADAR or bwDataDiss application environment will instantly allow researchers who use miscroscopes, cameras and other data taking instruments in e.g. chemistry, biology and genetics, to publish data directly from the workbench. In the third pillar of the project requirement and possibilities for data exchange between the repositories and publishers is being investigated. All components of bwDIM are tightly connected to a generic authentication and authorization infrastructure that governs and protects access to data. Work on the AAI components of bwDIM is done in close collaboration with the LSDMA and the European AARC [Ert16] project.

The bwDIM brings together researchers and developers from the KIT Library, the KIT computer centre (SCC) and the Leibniz Institute for Information Infrastructure (FIZ).

3 Other Users and Requirements

From the examples of use cases above it is clear that plain storage commonly provided by the data centre is not sufficient to cater for all the requirements of archive applications and usage.
There is a strong requirement for data checksums that are accessible at the user level. Archive applications can also benefit from the ability to be able to selectively keep certain data cached on disk and to be able to verify archived data without actually reading the data. Archive and long-time storage requirements were collected from communities participating in the Large Scale Data Management and Analysis [Jun14] initiative at SCC. The project brings diverse communities and several archive related projects together which resulted in a unique basis for discussion and assessments. More requirements stem from the infrastructure project bwDataArchiv [PW14], funded by the state of Baden- Württemberg, which aims to develop and deploy a service for long time storage for scientific data. It supports users from universities and institutes in Baden-Württemberg and national and international archive and large data preservation projects. The infrastructure and technology of bwDataArchiv offers a secure, reliable and tamper free storage for billions of files at PetaByte scale.

4 Archive System Technology

To securely store big data, tape systems have been the first choice at data centres for reasons of reliability and costs [RK13]. Maybe tape is the ugly sister of magnetic storage but with its longevity, reliability and the low, nearing zero power consumption of tape for storing digital data, it is clear that tape is a prime medium for archiving. Although disk density may further increase tape has a proven potential for high capacity storage. In the laboratory the tape recording layer with nano-grained particles reaches a density of 148 Gb/in^2 [16v], which is more than 30-40 times the density of today's tape

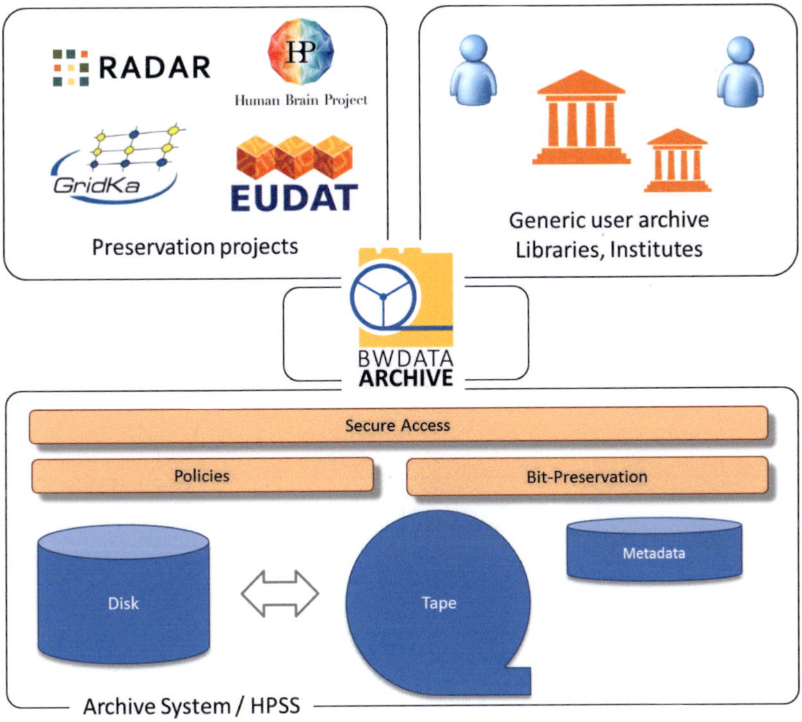

Figure 39: Institutions rely on the long time storage infrastructure from bwDataArchive.

media (but still only 20% of today's hard disks). The down side of tape storage is the long time-to-first-byte and difficult file management. However the high speed access and zero to none latency of disks and future silicon storage comes at a price.

Part of the inherent problems of linear storage can be mitigated by intelligent caching and automated migration of data using hierarchical storage management (HSM). Software that incorporates this functionality and handles large tape collections is for example the High Performance Storage System (HPSS) [17a] in use at SCC for the bwDataArchiv project [PW14] and at HLRS. HPSS is a scalable policy based HSM solution that can manage petabytes of data on disk as well as on tape [Wat05].

HPSS combines online disk cache and archive tape storage and allows defining multiple storage hierarchies and migration policies based on file size, access frequency and number of copies on tape. Support is available for end-to-end data integrity and validation using the IEEE T10-Protection Information field, which allows the checksum to be transferred from application to the host bus adapter such as a Fibre Channel card to the disk or tape drive and back. End-to-end data integrity protection prevents silent data corruption and theoretically guarantees a perfect bit to bit preservation. The recently introduced FUSE interface offers a file system like access to the HPSS tape storage, leveraging access through standard storage protocols like sftp, nfs, or http.

4.1 Archive Interface

In a collaboration that includes scientific users, data centre technicians and software developers the archive infrastructure at SCC will have offer enhancements for operating large data archives. The archive oriented interface, being developed, increases efficiency by enabling special operations for fixity checking, will allow storing of system meta data and informs the application about the location status of the stored objects. Ultimately the goal is to make archive storage interoperable between different applications and sites.
The variety of systems used for long term storage of data offer interface features that follow the POSIX standard to a certain degree and offer features to handle the low level operation of the system. Certain operations such as data validation are sometimes available but vastly different between products. Applications that use the storage i.e. dSpace , Preservica , Archivematica and many others, are therefore limited to the standard file system or database operations in case a data base is used for long time storage. Use of non-generic feature increases their complexity and makes them dependant on the underlying storage system. To cope with these problems a solution is developed

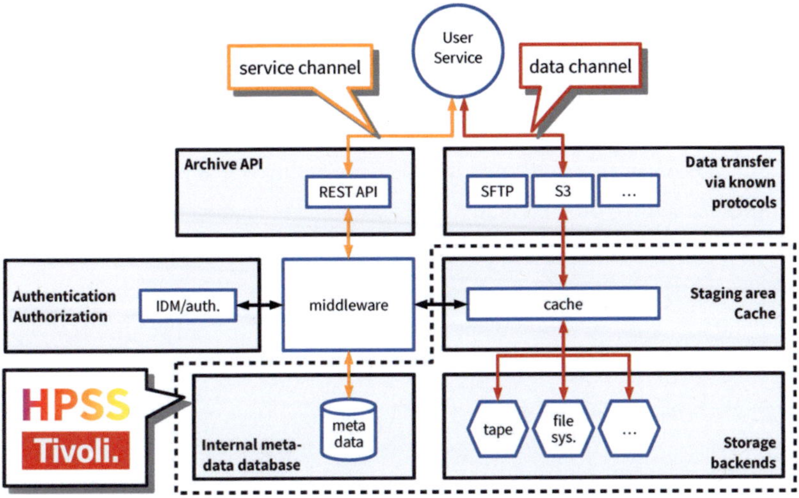

Figure 40: Components of the archive interface.

within the RADAR [16u] and EUDAT [16l] projects that is capable of unifying several tape storage systems.

4.2 Architecture

The system adds a service interface to the storage system alongside the common IO channel. While all of the IO operations are performed with the storage backend directly via the data channel, communication with our service is done on a separate channel, referred to as service channel (Figure 40).
This concept enables existing applications to store and access data in the usual way without interference with our service. If the application chooses to register data using the service channel it can use additional functionality. Currently the service enables applications to attach user defined attributes in the form of meta data and to access periodically calculated checksums.

5 Benefits to use cases and outlook

We have identified several communities and typical use cases that profit from storage with features dedicated to long time storage of data. Moreover, with the development of a dedicated interface for archives it is possible to use the storage backend more efficiently. The additional meta data layer which stores the checksum, can be used to register additional information that data centres require to operate long time storage for a growing number of communities. The implementation for efficient checksum calculation for two different tape based storage systems behind the same interface is a first step to the implementation of several features that will support archive applications and data centre operations. The uptake of the new interface in one or more projects at KIT helps to quickly improve the implementation and to selectively add features from user requirements.

Acknowlegdements

Presented work and projects receive funding by the Helmholtz Association, the state of Baden-Württemberg, the federal republic of Germany and the European Union.

Quoted LSDMA Publications

[Ert16] Benjamin Ertl et al. “Identity Harmonization for Federated HPC, Grid and Cloud Services”. In: *Proceedings of the 2016 International Conference on High Performance Computing and Simulation*. IEEE. 2016, pp. 621–627.

[Jun14] C Jung et al. “Optimization of data life cycles”. In: *Journal of Physics: Conference Series* 513.3 (2014), p. 032047. DOI: 10.1088/1742-6596/513/3/032047. URL: http://stacks.iop.org/1742-6596/513/i=3/a=032047.

Other References

[16e] *bwDataDiss project web site*. 2016. URL: http://www.alwr-bw.de/kooperationen/bwdatadiss/.

[16l] *EUDAT project web site*. http://eudat.eu/. [Online; accessed 10-October-2016]. 2016.

[16o] *German research foundation "Safeguarding Good Scientific Practice"*. Oct. 2016. URL: http://www.dfg.de/en/research%5C_funding/principles%5C_dfg%5C_funding/good%5C_scientific%5C_practice/.

[16t] *Open Data repository*. Oct. 2016. URL: http://www.open-access.net/.

[16u] *RADAR project web site*. Oct. 2016. URL: http://www.radar-projekt.org/display/RE/Home/.

[16v] *Sony news release on magenetic tape technology*. Oct. 2016. URL: http://www.sony.net/SonyInfo/News/Press/201404/14-044E/.

[17a] *HPSS Collaboration web site*. http://www.hpss-collaboration.org/. [Online; accessed 29-March-2017]. 2017.

[Gar11b] A.O. Garcia et al. *The Large Scale Data Facility: Data Intensive Computing for Scientific Experiments.* May 2011. DOI: 10.1109/IPDPS.2011.286.

[Hic10] Jason Hick. “HPSS in the Extreme Scale Era: Report to DOE Office of Science on HPSS in 2018-2022”. In: *Lawrence Berkeley National Laboratory* (2010).

[PW14] J. Potthof and J. van Wezel. “Landesprojekt bwDataArchiv – SCC erweitert digitales Archiv für Langzeitspeicherung von Forschungsdaten”. In: *SCC News* 2014-1 (2014), pp. 12–14.

[RK13] David Reine and Mike Kahn. “Revisiting the Search for Long-Term Storage—A TCO Analysis of Tape and Disk”. In: *The Clipper Group Calculator, May* 13 (2013).

[Wat05] Richard W Watson. “High performance storage system scalability: Architecture, implementation and experience”. In: *22nd IEEE/13th NASA Goddard Conference on Mass Storage Systems and Technologies (MSST’05).* IEEE. 2005, pp. 145–159.

Performance and Power Optimization

Michael Kuhn[a], Konstantinos Chasapis[a], Manuela Kuhn[b], Janusz Malka[b], Thomas Stibor[c], Gvozden Nešković[d]

[a] Universität Hamburg, Hamburg
[b] Deutsches Elektronen-Synchrotron DESY, Hamburg
[c] GSI Helmholtzzentrum für Schwerionenforschung GmbH, Darmstadt
[d] Frankfurt Institute for Advanced Studies, Frankfurt am Main

Abstract Managing and analyzing large amounts of data requires high performance storage systems that can keep up with the applications' I/O demands. Additionally, energy efficiency plays an important role as storage systems are often responsible for a significant part of the total cost of ownership. Within the Performance and Power Optimization work package of the Data Services Integration Team, we have focused on both of these aspects. Based on demands observed in real systems and applications, we have developed tools and solutions to improve both performance and cost efficiency.

1 Introduction

Throughout the history of supercomputers as recorded by the TOP500 list, the computational power has been increasing exponentially, doubling roughly every 14.5 months. This development is currently realized by employing an increasing number of processor cores, with the current number one system having more than 10 million cores. While this increase in computational power has allowed more detailed numerical simulations to be performed, this has also caused the simulation results to grow in size exponentially. Computational speed and storage capacity have roughly increased by factors of 300 and 100 every 10 years, respectively. The storage speed, however, has only grown by a factor of 20 every 10 years, even when taking newer technologies such as SSDs into account. Thus, the importance of performing I/O efficiently and storing the resulting data cost-efficiently increases.

Overall, the storage subsystems can be responsible for a significant portion of a system's total cost of ownership. In the case of Mistral, hosted at the German Climate Computing Center (DKRZ), it accounts for roughly 20 % of the overall costs. This amount of storage is necessary due to the huge amounts of data produced by current HPC applications. The I/O requirements of parallel applications, however, can vary widely: While some applications process large amounts of input data and produce relatively small results, others might work using a small set of input data and output large amounts of data; additionally, the data can be spread across many small files or be concentrated into few large files. Naturally, any combination thereof is also possible. These different requirements make high demands on supercomputers' storage systems.

To foster data exchange and portability, application developers often make use of high-level libraries that offer access to self-describing data. Middlewares and other abstractions are used to shield application developers from the complexity of HPC I/O. Additionally, parallel file systems themselves are split up into multiple layers. Combined, this leads to HPC I/O stacks as illustrated in Figure 41. While this allows exchanging individual layers without impacting others, this stack also has its downsides. On the one hand, this complex stack makes it important to adequately tune I/O for optimal performance because many different software components are involved until the data actually reaches the storage devices. On the other hand, debugging performance problems and optimizing I/O performance is made difficult by the complex interplay of layers.

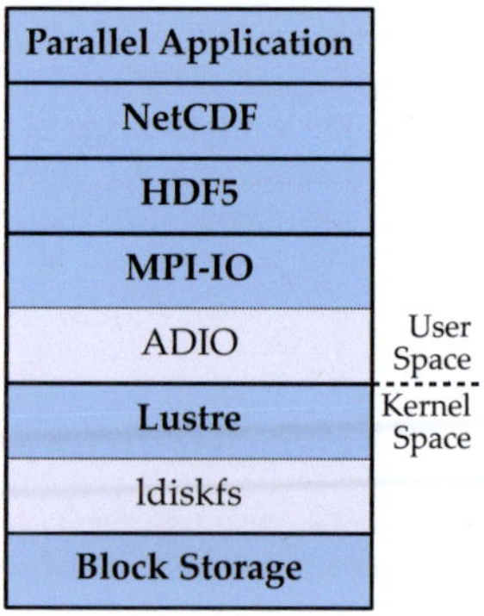

Figure 41: HPC I/O stack based on technologies typically used in earth system science.

Additionally, even though the theoretical computing performance continues to increase exponentially, applications have trouble scaling up to millions of processor cores. The actual utilization of often less than 10 % and it is therefore also necessary to improve application performance. The Performance and Power Optimization work package's mission has been to provide application developers with adequate tools and information on how to make use of the available HPC hardware in an efficient way, regarding both performance and energy. Additionally, important parts of the I/O infrastructure should be analyzed and optimized. The following sections are meant to give an overview of the most important achievements and illustrate their benefits for the wider data science community.

2 Metadata Performance in Lustre

File system metadata operations can be crucial to overall system performance, especially in the case that each application process operates on a single private file. In this scenario, the parallel file system is often hammered with a vast amount of metadata operations such as `create`, `unlink` or `stat`. While new algorithms and techniques to improve the metadata operation performance have been proposed, only little evaluation has been done with regards to newer multi-core and multi-socket non-uniform memory access (NUMA) platforms and the emerging storage devices such as solid state drives (SSDs) and non-volatile random access memory (NVRAM).

In [Cha14a], we examine the implications of such platforms regarding the performance scalability of Lustre's metadata server in version 2.4. We run our experiments on a four socket NUMA platform that has 48 cores. We leverage the `mdtest` benchmark to generate appropriate metadata workloads and include configurations with varying numbers of active cores and mount points. Additionally, we compare Lustre's metadata scalability with the local file systems ext4 and XFS.

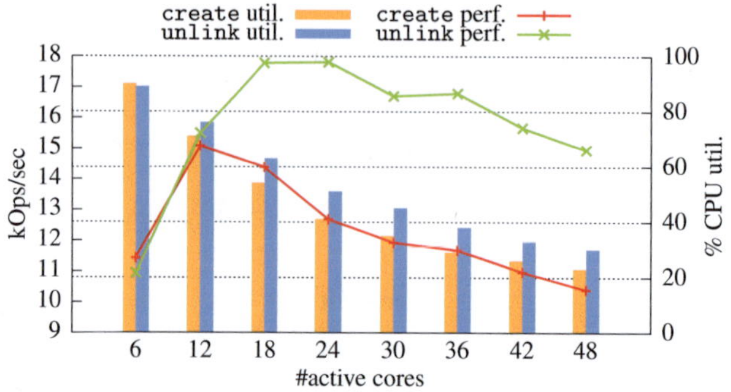

Figure 42: Lustre metadata performance and CPU utilization of the `create` and `unlink` operations with respect to the number of active cores.

Figure 42 illustrates the results for the `create` and `unlink` operations while varying the number of active cores in the system. As we can see, the performance improvement is limited to 18 cores for `create` and 12 cores for `unlink`. Also, we see huge performance degradations when using more cores. The results comply with the observed CPU utilization that drops when more cores are added. Running similar experiments with multiple configurations, we identify that the performance of Lustre's metadata server is limited to a single socket. We also observe that the metadata server's back-end device is not a limiting factor regarding the performance.

3 Performance Monitoring

Performance analysis is an important prerequisite for performance optimization. ElasticMonitoring is a DESY-developed monitoring solution based on ElasticSearch and uses Kibana/Grafana to navigate through the database and create a customized dashboard. The tool collects metrics and feeds an ElasticSearch database for further analysis (see Figure 43). It consists of three information sources: methods to collect performance metrics, syslog forwarding and customizable information collection. On the one hand, to collect performance metrics, IBM's ZIMon sensors and collectors for a GPFS-based

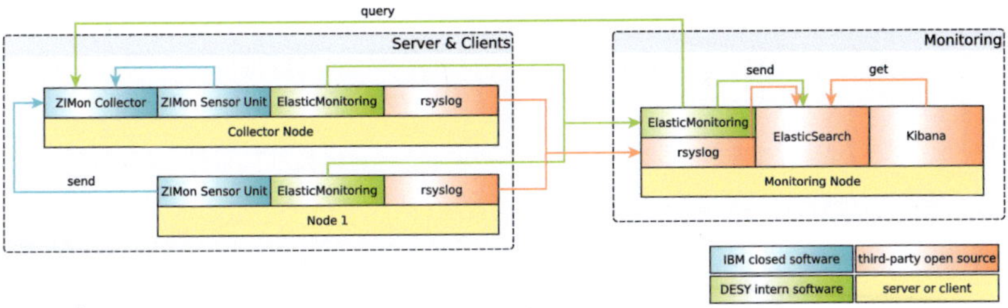

Figure 43: Architecture of the ElasticMonitoring suite.

system can be used as a data source. On the other hand, there is also the possibility to configure the metrics manually, making the tool independent from any commercial software. The latter can also work as an extension to the ZIMon method, for example, if additional metrics are needed that are not contained in the database. To collect system logs, the tool rsyslog is used together with an ElasticSearch plugin. Furthermore, a Python module was implemented to forward any customized information into the ElasticSearch database. All of this information together gives the possibility to correlate all collected data and events. This helps monitoring the resource usage, to find bottlenecks and to analyze and trace occurring errors. Additionally, it helps in planing of future resource needs. It is successfully used in DESY's newly created storage environment for the next generation of physics experiments at the local X-ray source Petra-III and their 24 beamlines [Str15].

3.1 Automatic Performance Validation Tool

For systems designed to have a continuous fast data ingest coming over different protocols, monitoring of the connections is mandatory. An example of such a system coming from photon science is one or multiple detectors taking images from a sample and writing it into a storage system. These detectors are running on different Windows or Linux type operating systems giving access to the data over a NFS or SMB connection. To detect problems, the whole data taking chain has to be validated. We developed an automatic per-

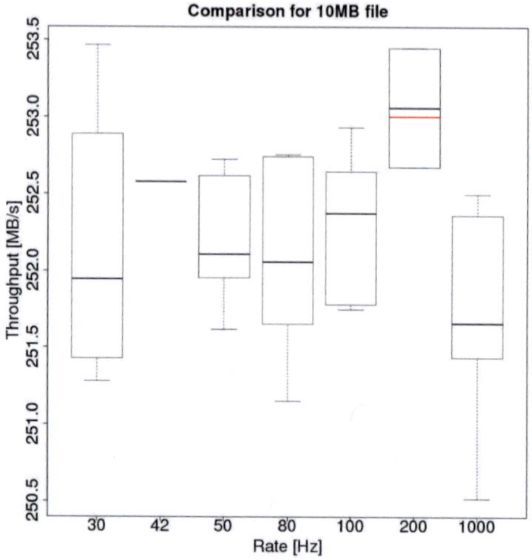

Figure 44: Deviations for a 10 MB test file.

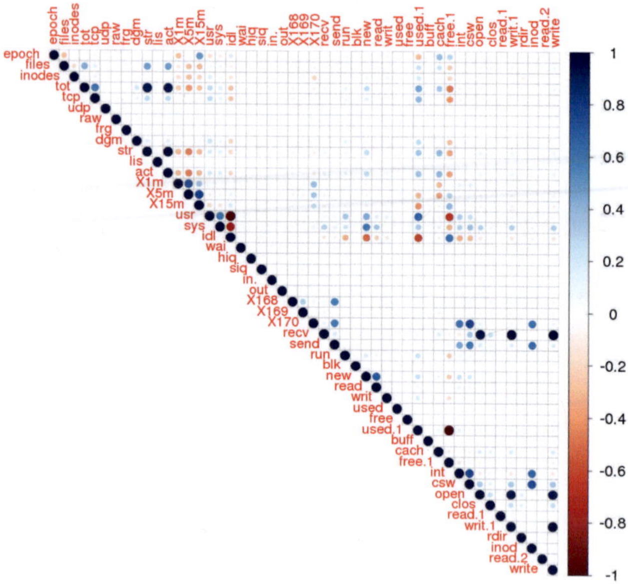

Figure 45: Correlation matrix.

formance validation tool to check NFS and SMB connections from a detector node to a storage system fully automatically. For that, the tool generates fake data with a given size and rate using a dedicated simulation tool and ingests it into the storage system. The measured performance results are then compared with a reference data set to find deviations (see Figure 44). Furthermore, different types of quantities are collected to find correlations and dependencies (see Figure 45). The tests are done on a regular basis to find potential problems as soon as possible.

4 Data Transfer and Online Analysis of Scientific Data

Current and future instrument development for scientific experiments reveal the need for tools to handle the huge amount of generated scientific data (see DLCL Matter). On the one hand, the data has to be drained from the detectors fast enough. On the other hand, the experimental conditions have to be monitored and analyzed in close to real time to prevent the collection of unfavorable data, which also helps with preserving the valuable sample.
HiDRA (High Data Rate Access) was developed exactly for that purpose (see Figure 46). It is a generic tool set for high performance data multiplexing with different qualities of service and is based on Python and ZeroMQ. It can be used to directly store the data in the storage system but also to send

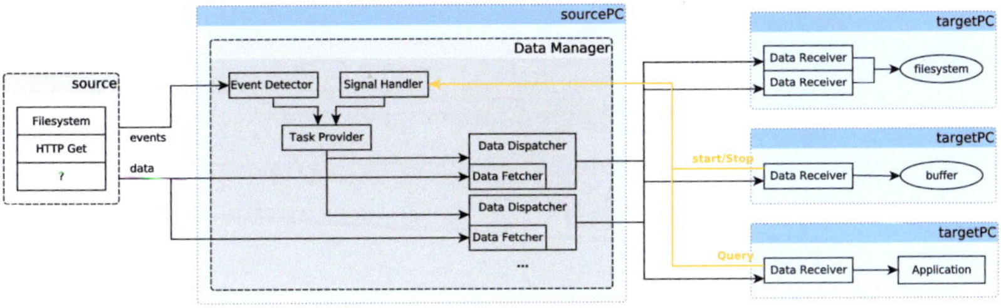

Figure 46: HiDRA architecture.

it to some kind of online monitoring or analysis framework. Together with OnDA (Online Data Analysis) [Mar16], data can be analyzed with a delay of seconds resulting in an increase of the quality of the generated scientific data by 20 %. The modular architecture of the tool (divided into event detectors, data fetchers and receivers) makes it easily extendible and even gives the possibility to adapt the software to specific detectors directly (for example, Eiger and Lambda detector).

Available event detectors:

- Based on inotifyx library (Linux)
- Based on watchdog library (Linux/Windows)
- Get events via API (C/Python)

Available data fetchers:

- Read from file system
- Get data via API (C/Python)

Available receiver types:

- Store as files
- Forward to an application
- Build HDF5

5 Best Practices for Scientific I/O with HDF and NetCDF

A typical example of I/O-intensive HPC applications are simulations from the field of earth system science that generate vast amounts of data as checkpoints and model output files. The performance of those applications is highly dependent on the storage system performance due to the synchronous nature of checkpointing. Another important aspect for such applications is efficient data management and, to improve this, standardized file formats are used to store and manipulate data so that they can be easily exchanged between institutes and scientists for further processing. Among the commonly used formats are the Hierarchical Data Format (HDF) and the Network Common Data Format (NetCDF). They are surrounded by APIs that allow manipulation and

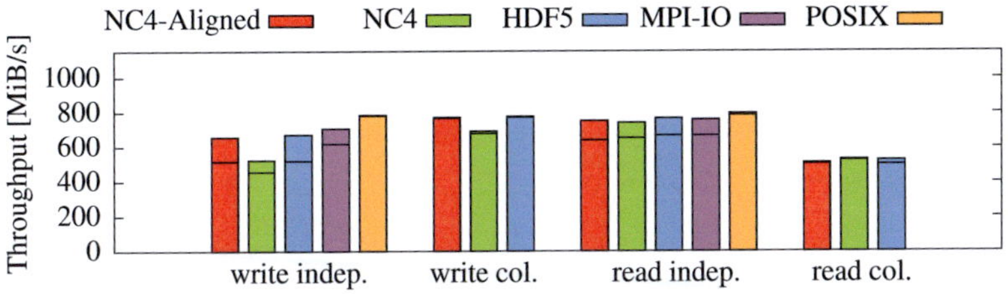

Figure 47: Performance with disjoint pattern.

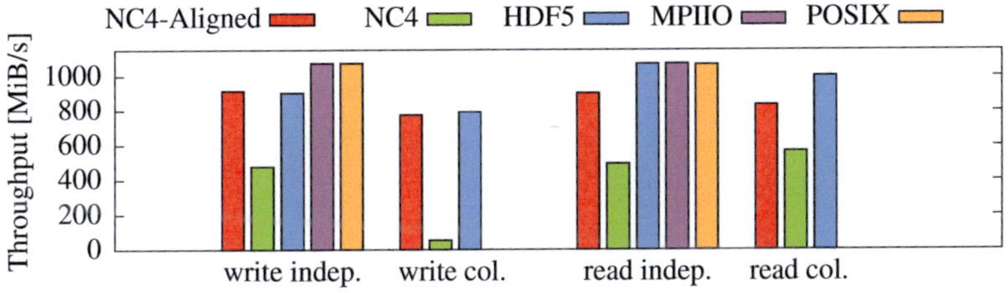

Figure 48: Performance with 1-OST pattern.

retrieval of the data by programs written in various programming languages like C or Fortran. Therefore, establishing best practices for developers using these high-level libraries is an important step in ensuring that available HPC hardware is well-utilized.

In [Bar15], we present these common I/O interfaces, evaluate their performance in detail and describe best practices to achieve optimal performance. To conduct our analysis, we used a self-modified version of the IOR benchmark to generate I/O patterns that mimic the I/O routines of real-world applications on top of Lustre due to its prevalence in HPC systems. While more results and best practices can be found in [Bar15], we want to illustrate the impact of access patterns on achievable performance. Figures 47 and 48 show results using a disjoint and 1-OST pattern, respectively. While the disjoint pattern causes all client processes to communicate with all file system servers, the 1-OST pattern establishes a 1-to-1 communication scheme between client processes and file system servers.

In Figure 47, many results have a high variation when using independent I/O due to a combination of competition on the file system servers and the network resources with the lack of synchronization when using independent I/O. The results are much lower than the practical maximum of 1,125 MiB/s. In Figure 48, POSIX and MPI-IO achieve almost the theoretical maximum because the 1-to-1 communication prevents competition on the servers and network resources. By default, NetCDF issues unaligned data accesses which causes significant performance degradations and leads to communication with more than one server. Enabling aligned accesses in HDF and implementing analogous functionality in NetCDF greatly improves performance and allows both libraries to almost reach the same performance as POSIX and MPI-IO.

6 Power Consumption Modeling

The continually increasing power needs for running a supercomputer set a barrier to their future development both for total cost of ownership but also the ability to provide the amount of energy required. The currently fastest supercomputer consumes approximately 15.3 MW to deliver 93 PFLOPS. In other words, if we continue with the same trend we will need a small nuclear power plant to generate enough electricity to power Exaflop supercomputers.

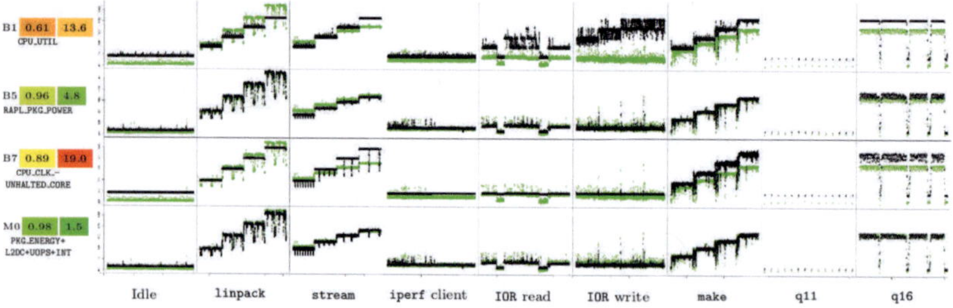

Figure 49: Real (green) and estimated (black) power consumption traces of some of the training and validation benchmarks. Note that the X-axis represents the timeline of samples while the Y-axis measures power from 38 to 160 W.

Monitoring the power consumption during an application run is an important step towards improving the energy efficient of it. The use of wattmeters is obviously the easiest solution. However, in most of the real case scenarios this is not feasible. To that end, the use of models to predict the power consumption of a system is an appealing alternative to wattmeters since they avoid hardware costs and are easy to deploy.
In the work presented in [Dol15], we illustrate an analytical methodology to build models with a reduced number of features in order to estimate power consumption at node level. We aim at building simple power models by performing a per-component analysis (CPU, memory, network, I/O) through the execution of four standard benchmarks. While they are executed, information from all the available hardware counters and resource utilization metrics provided by the system is collected. Based on correlations among the recorded metrics and their correlation with the instantaneous power, our methodology allows to identify the significant metrics and to assign weights to the selected metrics in order to derive reduced models. The reduction also aims at extracting models that are based on a set of hardware counters and utilization metrics that can be obtained simultaneously and, thus, can be gathered and computed on-line.
Figure 49 plots power traces in order to compare the real (green) and estimated (black) power consumption using some of the baseline cases, ranging from models that work with only one feature to more elaborated ones combining hardware counters, OS statistics and temperature sensors. We include some benchmarks (training set) but also cases from the validation set. We see that even though `CPU_UTIL` cannot predict the power consumption well, once we add more features we are able to minimize the difference between the predicted and actual measured value.

7 Scientific Data Compression

Due to the continuously increasing processing power in HPC systems and the slower pace of storage development mentioned previously, the I/O problems observed today will be much more prevalent in the future. Compression can be used to reduce the amount of data and thus address both performance and capacity problems.

7.1 Elaborate Compression

Compressing application level checkpoints, visualization dumps etc. efficiently is a challenging task since they are highly heterogeneous in nature. A single file usually contains values of variables of different data types and different representations such as single or multi-dimensional arrays. To this end, the efficiency of a single compression algorithm for the whole dataset is limited since it will not work best for all the cases. Moreover, domain decompositions of scientific data may introduce temporal and spatial locality within and across files containing scientific data. For example, in a climate simulation application, a variable modeling the temperature of Texas may have similar values to that of New Mexico. Current applications tend to use high-level file formats like HDF that allow data description (variable name, type, size, dimensionality etc.).
In the work presented in [Cha15b], we propose to improve scientific datasets compressibility by generating data clusters based on the knowledge provided by the high-level file formats and the similarity values and then applying adaptive compression per cluster based on the data patterns found on them. Our clustering technique aims to form discrete clusters with less entropy which will result in higher compressibility. The adaptive compression algorithm targets to exploit the data patterns found in each cluster by applying the appropriate compression algorithm.

7.2 Power/Performance Benefits

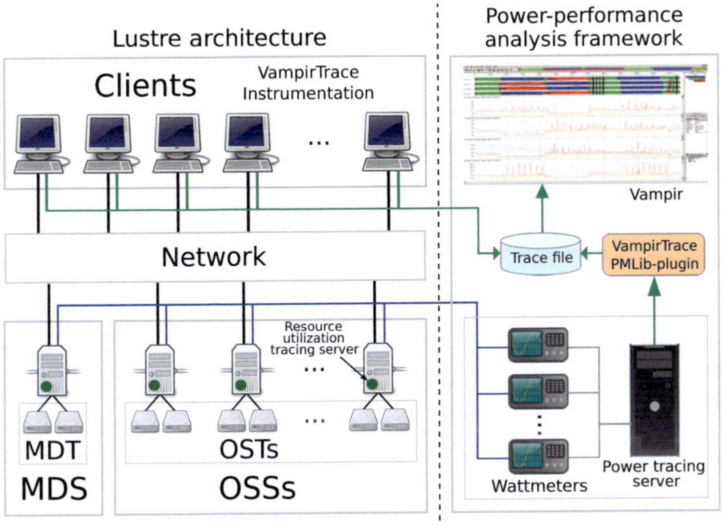

Figure 50: Lustre architecture in combination with the power-performance analysis framework.

In [Cha14b], we provide a power-performance evaluation of HPC storage servers that take over tasks other than simply storing the data to disk. We use the Lustre parallel file system with its ZFS back-end, which natively supports compression, to show that data compression can help to alleviate capacity and energy problems. In the first step of our analysis, we use the tracing environment shown in Figure 50 to study different compression algorithms with regards to their CPU and power overhead with a real dataset. Then, we use a modified version of the IOR benchmark to verify our claims for the HPC environment.

The results demonstrate that the energy consumption can be reduced by up to 30 % in the write phase of applications and up to 7 % for write-intensive applications. At the same time, the required storage capacity can be reduced by approximately 50 %. These savings can help in designing more power-efficient and leaner storage systems. Reducing the data size also leads to benefits in terms of total cost of ownership. In [Kuh14], we illustrate a model to estimate the benefits of compression in advance. The model only takes costs for the actual storage hardware into account but not possible indirect savings such as decreased cooling and space requirements.

8 Vectorization of Reed-Solomon Codes used in ZFS

ZFS uses RAID-Z technology to implement data protection. This is a software scheme that utilizes error correction coding (ECC) to minimize cost of mirroring in terms of required disks. Depending on selected scheme, RAID-Z1, RAID-Z2 or RAID-Z3, such a volume is able to recover from a single, two or three disk failures, respectively. We developed and implemented efficient scalar and SIMD vectorized versions of RAID-Z methods. Reconstruction of erasures in the complete RAID-Z scheme is implemented in seven specialized methods. The original implementation used a `Log/Exp` lookup table multiplication method that requires two lookup table operations per symbol to perform multiplication by a constant. For a 64-bit scalar implementation, this amounts to 16 lookup operations.
The impact of these optimizations is shown in Figure 51. For evaluation, RAID-Z block size is varied from 16 KiB to 64 MiB so that data does not fit into CPU caches. Plots show per-disk throughput of the original and new RAID-Z methods using scalar, SSE and AVX2 instruction sets, executing on a single CPU core. The evaluation assumes eight data disks and two parity disks (that is, RAID-Z2).

9 Analyzing and Predicting `LBUG`s with a Hidden Markov Model

Lustre is a parallel distributed network file system, employed predominantly in the domain of high performance computing. Due to the nature of such a

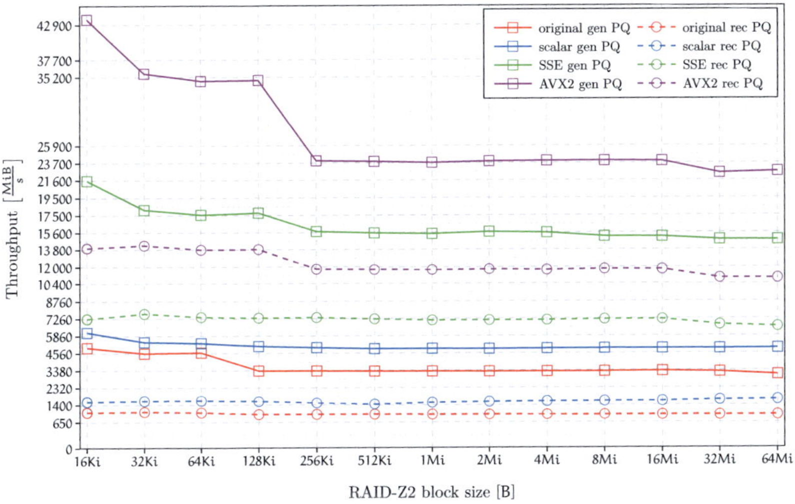

Figure 51: Combined throughput of RAID-Z2 parity operations on pool consisting of eight data and two parity disks.

parallel and distributed system and the large and complex code base, Lustre is prone to critical software bugs called LBUGs. These bugs cause freezing kernel threads and require a subsequent reboot of the operating system. We developed a hidden Markov model for analyzing and predicting LBUGs that is trained on Lustre logging data, which is a time series and consists of Lustre function calls and information whether LBUGs occurred. The model is trained with the Baum-Welch algorithm, that is, values in the hidden and emitting matrix are optimized such that the model can reproduce the input data in probabilistic sense. Moreover, hidden to emitting state, relations can be visualized for analyzing and understanding latent paths of functions calls (see Figure 52). These latent paths can reveal hidden relations of function calls that result in LBUGs.

In addition, one can sample, that is, generate data from the model. Furthermore, one can estimate the expected value of the time step a LBUG will occur and thus predicting time windows with LBUG occurrences.

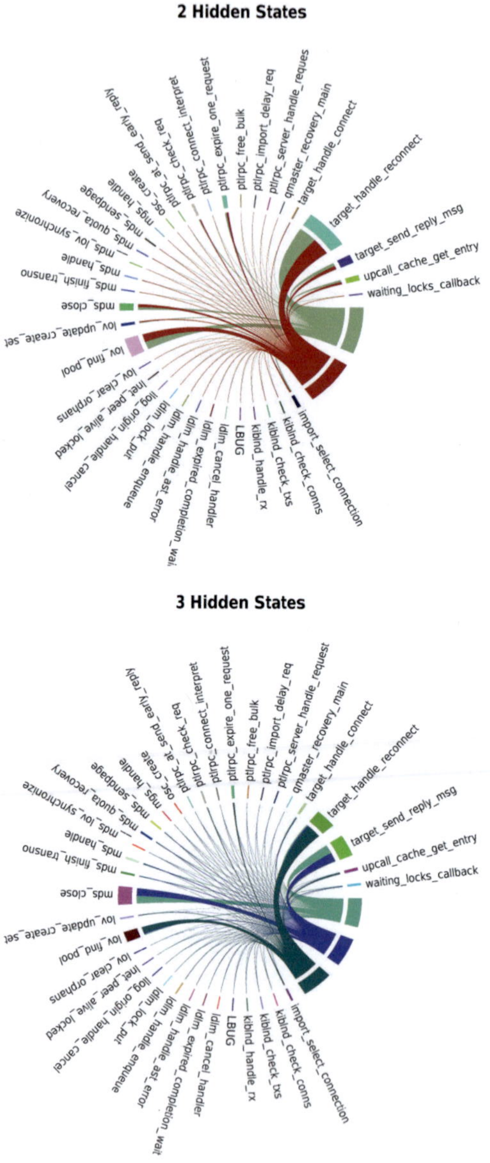

Figure 52: Visualization of latent paths from hidden to emitting states of Lustre logging data.

10 Conclusion and Contributions

The presented works cover a wide range of applications and use cases. The members of the Performance and Power Optimization work package have contributed both to demands of individual projects and to general components that are important for the storage infrastructure. To summarize, we have produced the following contributions that benefit data scientists and the whole data science community:

- Our metadata evaluations allow tuning storage system appropriately for workloads involving many small files and high metadata rates in general.
- A scalable solution for performance monitoring has been developed and deployed at DESY, which improves productivity of scientists by finding problems early.
- The online analysis of experiment results gives scientists immediate access to their data and allows them to tune their experiments appropriately.
- Best practices for HDF and NetCDF give hints to application developers using these and similar high-level libraries for achieving optimal I/O performance.
- Investigating compression in the HPC context has provided interesting insights regarding performance and power benefits. It has also led to several follow-up projects such as adding compression support to Lustre and adaptive compression to HDF.
- The improved performance gained by vectorizing the error correction coding in ZFS is beneficial for common configurations and also important in the HPC context as Lustre is moving towards using ZFS as its back-end file system.

- Analyzing and predicting LBUGs in Lustre can be used to improve the availability of the file system. This is especially important as LBUGs typically require rebooting the affected machines, leading to extended downtime.

- Additional cooperations and studies based on work done in this work package that could not be mentioned in this article due to space reasons [LAK15; KKL14].

Acknowledgement

We thank everyone involved in the publications and work associated with this work package.

Quoted LSDMA Publications

[Cha14b] Konstantinos Chasapis et al. "Evaluating Power-Performace Benefits of Data Compression in HPC Storage Servers". In: *IARIA Conference*. Ed. by Steffen Fries and Petre Dini. Chamonix, France: IARIA XPS Press, Apr. 2014, pp. 29–34. ISBN: 978-1-61208-332-2.

[KKL14] Julian Kunkel, Michael Kuhn, and Thomas Ludwig. "Exascale Storage Systems – An Analytical Study of Expenses". In: *Supercomputing Frontiers and Innovations*. Volume 1, Number 1 (June 2014). Ed. by Jack Dongarra and Vladimir Voevodin, pp. 116–134. URL: http://superfri.org/superfri/article/view/20.

[Kuh14] Michael Kuhn et al. "Compression By Default - Reducing Total Cost of Ownership of Storage Systems". In: *Supercomputing*. Ed. by Julian Martin Kunkel, Thomas Ludwig, and Hans Werner Meuer. Lecture Notes in Computer Science 8488. Leipzig, Germany: Springer International Publishing, June 2014. ISBN: 978-3-319-07517-4. DOI: 10.1007/978-3-319-07518-1.

[LAK15] Michael Lautenschlager, Panagiotis Adamidis, and Michael Kuhn. "Big Data Research at DKRZ – Climate Model Data Production Workflow". In: *Big Data and High Performance Computing*. 26th ed. Advances in Parallel Computing. IOS Press, 2015, pp. 133–155. ISBN: 978-1-61499-582-1. DOI: 10.3233/978-1-61499-583-8-133. URL: http://ebooks.iospress.nl/volume/big-data-and-high-performance-computing.

[Str15] M. Strutz et al. "ASAP3 - New Data Taking and Analysis Infrastructure for PETRA III". In: *J. Phys. Conf. Ser.* 664.4 (2015), p. 042053. DOI: 10.1088/1742-6596/664/4/042053.

Other References

[Bar15] Christopher Bartz et al. "A Best Practice Analysis of HDF5 and NetCDF-4 Using Lustre". In: *High Performance Computing*. Ed. by Julian Martin Kunkel and Thomas Ludwig. Lecture Notes in Computer Science 9137. Frankfurt, Germany: Springer International Publishing, June 2015, pp. 274–281. ISBN: 978-3-319-20118-4. DOI: 10.1007/978-3-319-20119-1_20.

[Cha14a] Konstantinos Chasapis et al. "Evaluating Lustre's Metadata Server on a Multi-socket Platform". In: *Proceedings of the 9th Parallel Data Storage Workshop*. PDSW 2014. New Orleans, Louisiana: IEEE Press, 2014, pp. 13–18. ISBN: 978-1-4799-7025-4. DOI: 10.1109/PDSW.2014.5.

[Cha15b] Konstantinos Chasapis et al. *Towards Scientific-Data Compression Using Variable Clustering*. Livermore, California, Aug. 2015.

[Dol15] Manuel F. Dolz et al. "An analytical methodology to derive power models based on hardware and software metrics". In: *Computer Science - Research and Development* (2015), pp. 1–10. ISSN: 1865-2042. DOI: 10.1007/s00450-015-0298-8.

[Mar16] Valerio Mariani et al. "OnDA: online data analysis and feedback for serial X-ray imaging". In: *Journal of Applied Crystallography* 49.3 (June 2016), pp. 1073–1080. DOI: 10.1107/S1600576716007469.

Big Data and Realtime Computing

Hermann Heßling[a]

[a] University of Applied Sciences, HTW Berlin, Berlin

Abstract The resolution power of experimental devices in almost any area is increasing steadily resulting in data rates so huge that only some small fraction of the data can be stored in long–term data archives. Consequently, techniques have to be developed for an effective pre–analysis already during the data taking period in order to weed out that part of the data that is dispensable for an in–depth analysis later on. In this article, recent efforts in this direction are reviewed for photon science and radio astronomy.

1 Introduction

Photon science and radio astronomy are similar both from a physical point of view and from a computational point of view. This is particularly evident in the forthcoming new experiments European XFEL and Square Kilometre Array (SKA).

Computationally, both experiments are going to produce large amounts of data, so huge that only a small fraction of the data can be stored in long–term archives for a detailed scientific analysis. Already during the data taking period, algorithms have to perform a preliminary analysis in order to identify effectively data of interest in realtime or near–realtime. This is a strong challenge in photon science and, possibly, the most critical challenge at SKA.

Physically, both experiments perform interference measurements, i.e. they crucially make use of the superposition property of electromagnetic waves. In order to recover the information about the objects of interest, for example pulsars in radio astronomy and molecules in photon science, an inverse operation has to be applied to the measurement data which, in both experiments, is essentially an inverse Fourier transformation. In radio astronomy, the inverse Fourier transformation allows to determine the position of astronomical objects on the celestial sphere, i.e. the direction from which the radiation is emitted. The third dimension cannot be reconstructed in radio astronomy, in general, i.e. the relative distance between astronomical objects is not resolvable. In photon science, the diffraction images collected by the detectors are also two–dimensional. However, three–dimensional models of the samples can be reconstructed by extracting phase informations from the diffraction images which, in turn, is feasible as the orientation of the samples can take any value relative to the direction of the incoming laser light.

Despite the similarities there are also conceptual differences between XFEL and SKA which, in particular, have an impact on the problem where the reduction of data can take place.

In photon science, the relevant interference phenomena happened already when the laser light leaves an illuminated sample, i.e. a pre–selection of data can be realized as an on–detector vetoing, in principle. The experiments in photon science are not as fixed as the experiments at the Large Hadron Collider (LHC) at CERN but change dynamically in the course of time. Consequently, data reduction in real–time has to be automated in a way which is compatible with the flexibile setup of the experiments in photon science.

In radio astronomy, the individual detectors, i.e. the antennas, cannot be used for a pre–selection of data as the interference takes place later on within the "correlators" where the signals of any two antennas of an array of antennas are superimposed. See Sec. 3 for more details on the workflow at SKA.

2 Photon Science

In photon science, the internal structures of tiny objects like molecules, proteins, and viruses are explored by illuminating samples of such objects with laser light and collecting the diffracted light by means of detectors equipped

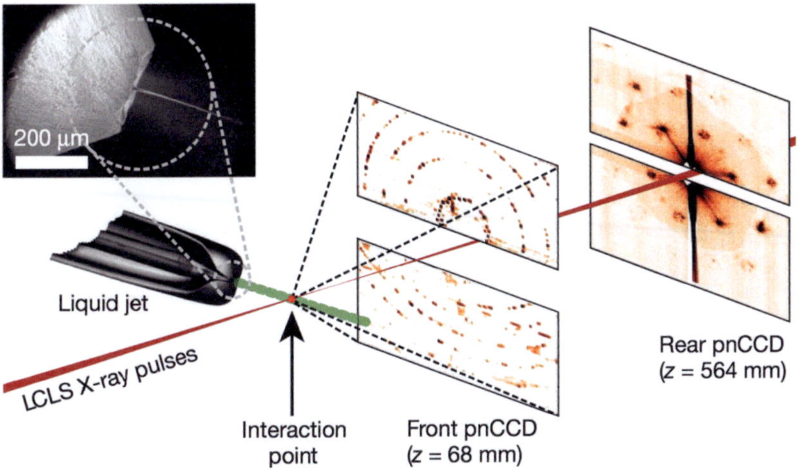

Figure 53: In femtosecond X–ray nanocrystallography, ultrashort light pulses from a free electron laser hit tiny crystals at the interaction point. The samples are transported in a stream which is propagated perpendicular through the laser beam. At the LCLS, the detector consists of two 2D–panels (each equipped with a 1024 x 1024 pixel array) and records coherent diffraction images at the rate of up to 200 Hz [N11].

with arrays of pixel sensors. In Fig. 53 the schematic layout of the Lineac Coherent Light Source (LCLS) is shown which is typical for experiments in photon science. In femtosecond nano–crystallography, crystals are formed out of the samples which have spatial extent of the order of nanometers and which are transported through a beam of light. The beam of light is composed of flashes of ultra–short X–rays where the the temporal extension of the flashes is of the order of femtoseconds which is briefer than the timescale where samples are destroyed by radiation damage. By this approach, the structure of macromolecules can be determined that cannot be combined to sufficiently large crystals necessary for studies using conventional radiation sources. A typical image from a macromolecule is shown in Fig. 54. The sharp dark peaks are called Bragg–spots and are due to a scattering of light at the internal crystal planes of the sample.

Figure 54: Diffraction image from a single nanocrystal built from the protein lysozyme [AHC14]. The distribution of the rectangles corresponds to the location of the sub–panel within the front panel at the LCLS. The dark spots show the Bragg–spots. The photon background is given by the "halo ring" around the beamhole at the centre of the detector.

The samples are dissolved in a transport medium, e.g. water, where they can move freely. This setup has an advantage and a disadvantage at the same time. The orientation of the nanocrystals is not fixed, i.e. a 2D–diffraction image taken by the detector is nothing but a projection along some random direction. This indeterminism can be used advantageously as the orientation of the sample can be reconstructed for every 2D image by comparing sufficiently many images. Finally, by merging all 2D–images properly a 3D–diffraction image can be obtained, see Fig. 55.[1] This information is used for extracting the electron density map of the sample, see. Fig. 56, essentially by performing an inverse Fourier transformation.

[1] For reconstructing 3D–structures it is not sufficient to consider the distribution of Bragg–spots, in addition, relative phases have to be extracted ("phase problem"). This can be done by exploring small regions around the Bragg–spots and by comparing their shapes with the shape obtained by averaging over samples whose Bragg–spots are located at the same position in the diffraction images. This method provides good results even if the signal–to–noise ratio (SNR) is low, see Ref. [CP12].

Figure 55: Three-dimensional rendering of the X-ray diffraction pattern for the Photosystem I protein, reconstructed from more than 15,000 single nanocrystal snapshots taken at the LCLS. Image: T. White, DESY.

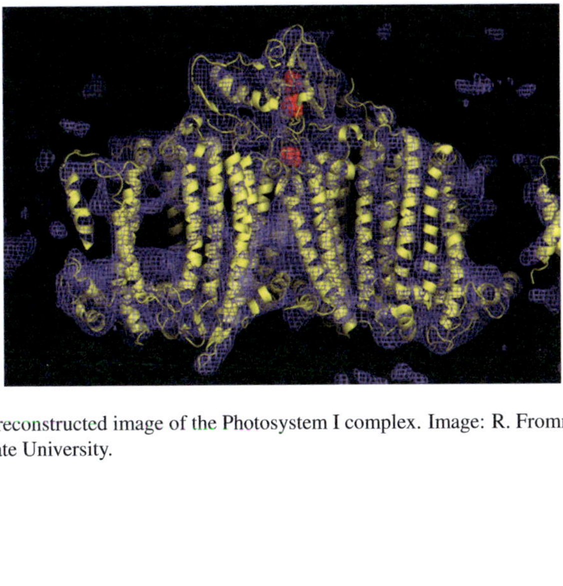

Figure 56: A reconstructed image of the Photosystem I complex. Image: R. Fromme, Arizona State University.

Within femtosecond nanocrystallography, it is not possible to synchronize the position of the samples and the time sequence of flashes such that every flash of X–rays hits a sample. On the contrary, the hit rate is quite small, only of the order of 5 %, i.e. most of the diffraction images taken by the detector are "blank".

Traditionally, every image collected by the experiments in photon science is stored and analyzed later on. This strategy could be realized because the data rates are comparatively small in previous and current experiments and because of the fear to loose valuable data due to an ineffective pre–selection of data.

With the upcoming European XFEL this traditional strategy is not feasible anymore and a shift in paradigm is needed. An identification of "blank" images should take place as early as possible preferable already within the detector and only "hit" images should be stored for a later analysis. The essential question is whether a fast algorithm can be found that is able to weed out successfully "blank" images in realtime. The available time period is given by the 27,000 images per second that can be taken at the European XFEL.

This problem was analyzed in a series of publications [BS14], [BS15a], [BS15b], [BS16a], and [BS16b], which are briefly reviewed in the following.

Large neural networks are successfully used for classifying images. However, the first attempt for using neural networks for identifying "blank" diffraction images failed [D16]. A very large three–layered neural network was constructed where the pixels of the 1024×1024 pixels of the front panel (see Fig. 53) were identified with the neurons of the input layer of the neural network, its hidden layer had the same number of neurons as the input layer, and the output layer showed two neurons. The training of this neural network turned out very time–consuming and, more importantly, the percentage of successfully identified "blank" images was approx. 50 %, i.e. this large "brute–force" neural network turned out to be blind and was just as effective as tossing a coin. The deeper reason for this failure is that the physical content

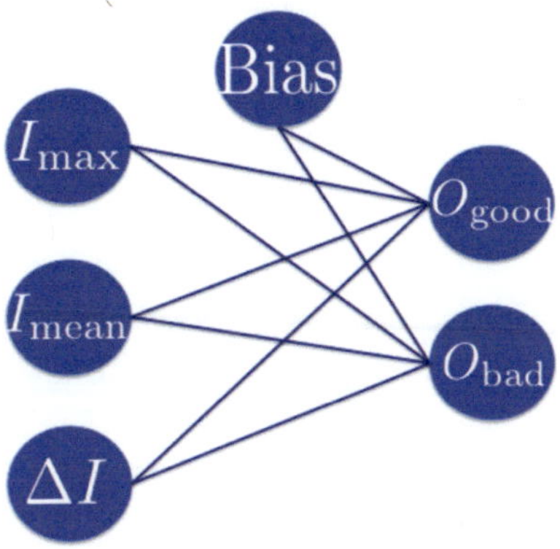

Figure 57: Neural network with three input neurons, two output neurons and no hidden layer. The weights of the connections between the neurons are obtained from a training set of diffraction images with known properties (either "hit" or "blank").

of the diffraction images is contained in the Bragg–spots, i.e. in rather point–like objects which are beyond the detection horizon of traditional imaging algorithms and even convolutional neural networks.

The decisive stimulation for deriving a solution was given by radio astronomy where neural networks successfully support the search for pulsars [WG10]. Remarkably, the neural network is quite small and consists of only twelve input neurons, twelve hidden neurons and two output neurons. The key to this approach is the extraction of twelve characteristic observables from a time–series analysis of the data and the identification of the observables with the neurons of the input layer.

Is it possible to implement this strategy also in photon science? In other words, are there characteristic observables in photon science that can be used as the input of a neural network? The answer is yes as is argued in Ref. [BS14]. Surprisingly, only three observables are sufficient and the complexity of the neural network is almost trivial as no hidden layer is needed, see Fig. 57. The observable I_{max} denotes the maximum intensity measured by one of the pixels, I_{mean} is the mean intensity obtained by averaging over all pixels and ΔI is the associated standard deviation. The value of the bias neuron is always set to one, as usual. A diffraction image is classified as "blank" if the output neurons obey the inequality $O_{good} \leq O_{bad}$. If the signal–to–noise ratio (SNR) of a sample of nanocrystals is sufficiently large, the diffraction images are almost correctly identified as "blank" or "not blank" (up to 95 %). In the subsequent papers [BS15a], [BS15b], and [BS16a] the algorithm was refined by removing single–pixel noise, see Fig. 59, by determining the boundaries of the Bragg–spots, see Fig. 60, and finally by identifying the positions of the Bragg–spots signals within a diffraction image, see Fig. 61. It turns out that the algorithm is identifying signals even within the "halo–ring" which is due to strong noise from diffraction of X–rays at the transport fluid of the samples. The reference tool Cheetah [AHC14] is less successful in identifying Bragg–spots within the "halo–ring".

In Ref. [BS16a] a workflow is suggested for implementing the neural network based algorithm (as shown in Fig. 58–61) into the detector hardware and it was estimated that an effective identification of "blank" diffraction images can successfully be realized in realtime at the European XFEL.

3 Radio Astronomy

In radio astronomy, electromagnetic radiation emitted from far–distant astronomical objects like pulsars, white dwarfs, and galaxies is measured by antenna arrays, see Fig. 62. The signals are interfered within the *c*orrelator by superimposing the collected signals between every pair of antennas. As a result, a complex–valued *v*isibility function $V(u,v)$ is obtained. Each pair of antennas corresponds to a point in the *uv*–plane. The coordinates (u,v) span a plane orthogonal to the direction of the astronomical object and are proportional to the *b*aseline between two antennas, i.e. the relative distance between both antennas.

As an example, Fig. 63 shows the visibility function of a double radio source system (e.g. radio jets or hotspots). The geometry of the system is encoded in the pattern structure of light and dark regions ("fringes") and can be made

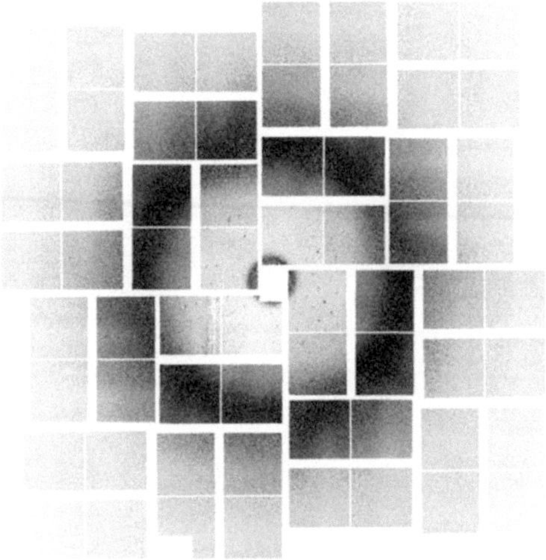

Figure 58: Algorithm: initial image [D16].

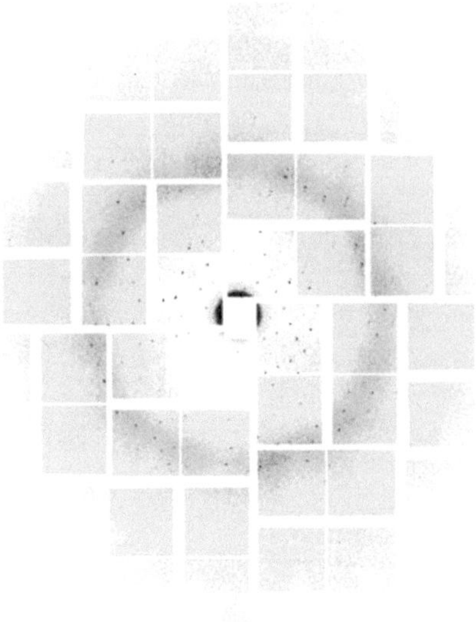

Figure 59: Algorithm: single–pixel noise removed [D16].

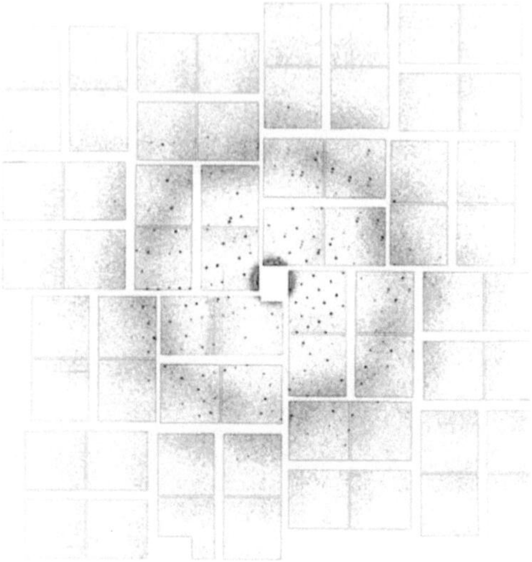

Figure 60: Algorithm: edge detection [D16].

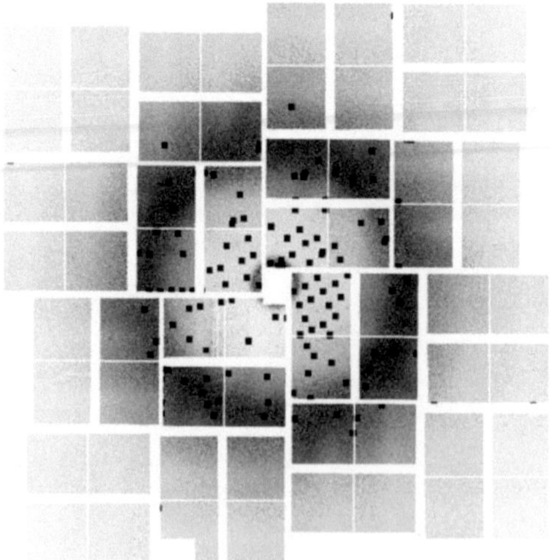

Figure 61: Algorithm: position of identified signals within the initial image [D16].

Figure 62: SKA dishes in South Africa (artists rendition, SKA organization). The 15m wide dish telescopes will provide highest–resolution imaging capabilities.

visible by an inverse Fourier transformation. The resulting 2D intensity distribution of the double–source (on the celestial sphere) is shown on the lower left corner of Fig. 63. Both sources are in contact with one another, have the same diameter, and are emitting radiation with the same intensity.
The larger the number of antennas the more points in the *uv*–plane are covered. Due to the rotation of the earth the position of each pair of antennas is changing resulting in trajectories in the *uv*–plane which is improving the overall covering and is essential for a determination of "fringes".
SKA is realized in two steps. In phase 1, approx. 10 % of the antennas are installed, the remaining 90 % are built in phase II. The amount of raw data collected by the antennas is expected to be $\sim$2 petabyte per second once SKA is fully operational, i.e. SKA and the Large Hadron Collider (LHC) will be comparable with respect to the raw data production rate. At LHC the raw data are reduced strongly by a factor $\sim 20^6$ in realtime and near–realtime based on multi–level trigger analyses, resulting finally in an increase of long–term data archives by $\sim$25 petabyte per year.

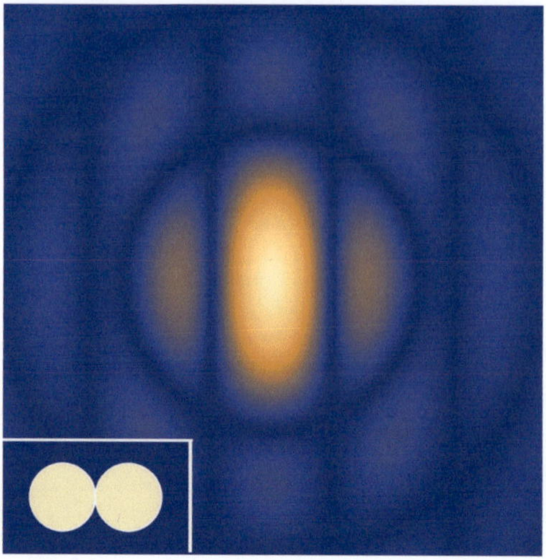

Figure 63: Visibilty function (absolute value thereof) of a double radio source system in contact with one another (see lower left corner). The circular fringes are due to the circular shapes of the sources. From the orientation of the straight fringes the orientation of the system can be extracted (the straight fringes are perpendicular to the axis between both radio sources). The boundaries of the fringes are given by the zeroes of the visibility function.

Data reduction in radio astronomy is much harder than at LHC. The data workflow at SKA is shown in Fig. 64. Data from the correlators are transported to one of two SKA Science Data Centers (located in Australia and South Africa, respectively). After meta data have been added data reduction takes place in two separated sub–workflows. By repeatedly processing the visibility function and by taking additional information into count (calibration, sky models, etc.) the amount of data is reduced and eventually standardized *d*ata products are formed which facilitate reprocessing and scientific analyses and which are delivered to worldwide–distributed *S*KA Regional Centers. In a second processing pipeline, time–series analyses are performed, for example for studying pulsars, and the resulting data products are also transported to the SKA Regional Centers. There, the data are stored

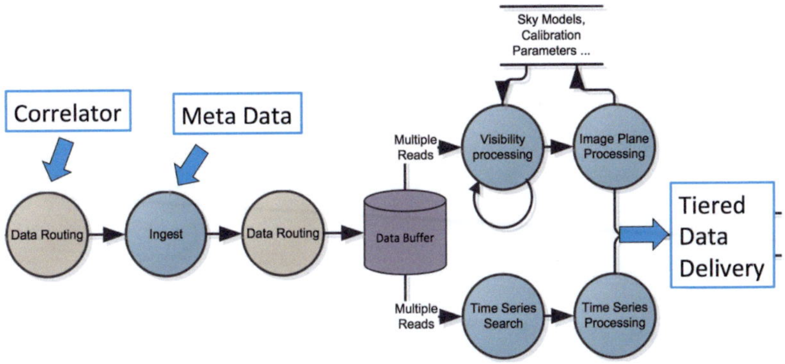

Figure 64: Data flow at the Square Kilometre Array (SKA) [P16].

in long–term data archives and can be processed by astronomers for physical analyses.

Within the Science Data Centers pre–analyses are performed in order to reduce the data by a factor $\sim 20^4$, i.e. at SKA one hundred times more data have to be stored in long-term data archives than at LHC due to the larger complexity of data in radio astronomy compared to high energy physics. In phase I, the Science Data Centers deliver up to one petabyte per day to the Regional Centers. Possible strategies for managing the huge data volumes are still being determined, for an overview of the current status see Ref. [P16]. It is suggested to generalize the popular MapReduce ansatz to graph–based data workflows. In order not to limit scaling, the movement of data and the exchange of messages is minimized. At SKA phase II, the necessary processing power and the size of the long–term data archives will increase by a factor of ~ 200.

The basic data products provided by the SKA Science Data Centers are "3D image cubes" whose sizes can be estimated from the final baseline design [al15]. For example, an image cube from the SKA–mid antennas contains data of approx. $10\text{k} \times 20\text{k}$ pixels, 4 polarisations, and ~ 64k frequency channels, i.e. a spectral image cube may be as large as 200 terabyte (provided

64 bits are stored per pixel). Data objects of this enormous size can no longer be processed by single workstations but need enormous parallel processing capabilities.

The big data challenges at SKA are so huge that new algorithms and tools have to be developed. The unavoidable parallel processing of large image cubes may require an exchange of intermediate results between parallel instances. This could have a strong impact on the scaling behavior of the workflow because, according to Amdahl's law, the speedup may go down due to blocking effects if a lot of messages of large size are exchanged between too many parallel instances.

How shall the long–term data archives be distributed between the different areas of research? This was not a problem before because all data were archived and could be analyzed in–depth later on. At SKA, however, the Science Data Centers perform fast pre–analyses, i.e. the scientific content of data products cannot be determined completely. To cope with the resulting uncertainty, probabilities should be assigned to data products indicating their relevance for scientific subareas. In high energy physics, in particular at LHC, almost any physical analysis relies on a full Monte Carlo simulation including the detector. Similarly, a full Monte Carlo simulation should be realized at SKA for reliably assigning probabilities to the data products. This requires an identification of observables that can be used for separating scientific subareas and that can be extracted out of the data in realtime [H16].

4 Contributions and Outlook

Only a small fraction of the data produced at the upcoming new experiments in photon science and radio astronomy can be stored in long–term data archives. For the European XFEL, an algorithm was developed to reduce the data volume by weeding out diffraction images that contain no physical information. Its essential component is a surprisingly small neural network whose input is identified with three observables that are determined by pre–

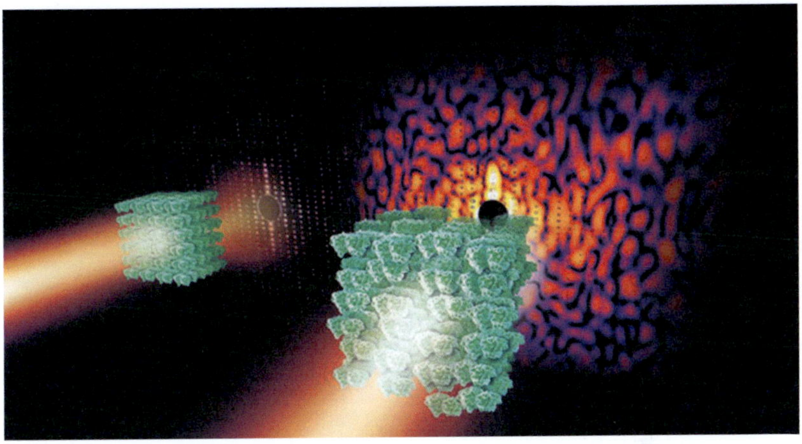

Figure 65: Diffraction images from weakly unordered crystals of giant biomolecules are blurred. Here, a diffraction image of the Photosystem II is shown. Out of the impressive patterns more information about the samples can be extracted than from diffraction images with Bragg–spots generated out of regular crystals (left). Image: E. Reimann, DESY.

analyzing a given diffraction image. It is suggested to run the algorithm in parallel within the panels of the detector hardware in order to identify useless "blank" diffraction images in realtime already during the data taking period. The proposed algorithm is an impressive example of a successful cooperation between informatics and physics.

The algorithm is shown to be quite effective if the diffraction images are taken from samples of regular nanocrystals. Forming regular nanocrystals out of organic giants like proteins and viruses is hard if possible at all. Remarkably, the internal structure of samples may be resolved even better if they are included in weakly irregular nanocrystals [al16]. The resulting diffraction images are blurred, see Fig. 65. Since the proposed algorithm relies on sufficiently strong signal–to–noise ratios, its effectiveness needs to be reinvestigated for this new class of samples.

Finally, the problem of taming the flood of data at the Square Kilometre Array (SKA) was considered. It was pointed out that for an effective reduction of data it may be beneficial to develop a Monte Carlo program for simulating both the radiation from astronomical objects and the influence of antennas. An open question is whether observables can be identified that allow to separate effectively and efficiently the different areas of research at SKA in realtime.

Acknowledgments

Fruitful discussions with the members of the DSIT group are gratefully acknowledged, in particular with Daniel Becker, Patrick Fuhrmann, Martin Gasthuber, Christopher Jung, Yves Kemp, Paul Millar, and Marco Strutz. Advise and comments given by Steve Aplin, Anton Barty, and Henry Chapman on photon science, and by Hans–Rainer Klöckner, and Michael Kramer on radio astronomy were most invaluable. It is a great pleasure to thank Volker Gülzow, and Achim Streit for their strong support during the LSDMA project.

Quoted LSDMA Publications

[BS14] D. Becker and A. Streit. "A neural network based pre–selection of big data in photon science". In: BDCloud. 2014, pp. 71–76. DOI: 10.1109/BDCloud.2014.42.

[BS15a] D. Becker and A. Streit. *Localization of signal peaks in photon science imaging*. Tech. rep. UKSim, 2015. DOI: 10.1109/uksim.2015.35.

[BS15b] D. Becker and A. Streit. "Real–time signal identification in big data streams". In: *Bragg–spot localization in photon science*. 2015, pp. 611–616. DOI: 10.1109/HPCSim.2015.723710.

[BS16a] D Becker and A Streit. "Real-time Signal identification in Photon Science Imaging". In: *IJSSST* (2016).

[BS16b] D. Becker and A. Streit. "Realtime–Processing of Nanocrystallography Images". In: UKSim-AMSS. 2016. DOI: 10.1109/UKSim.2016.20.

[H16] Heßling H. *Monte Carlo pathfinding in radio astronomy*. Tech. rep. GLOWSKA, 2016.

Other References

[AHC14] Barty A Kirian R A Maia F R N C Hantke M Yoon C H White T A, Chapman H, and Cheetah. "Software for high-throughput reduction and analysis of serial femtosecond X-ray diffraction data". In: *Journal of Applied Crystallography* (2014): *47, 3*, pp. 1118–1131. DOI: 10.1107/S1600576714007626.

[al15] Dewdney P et al. *SKA1 system baseline v2 description. Rev 01*. Tech. rep. Science and Technology Facilities Council, 2015.

[al16] Ayyer K et al. "Macromolecular diffractive imaging using imperfect crystals". In: *Nature* (2016): *16949*.

[CP12] Chen J P Spence J C and Millane R P. "Phase retrieval in femtosecond X–ray nanocrystallography". In: *Proceedings of the 27th Conference on Image and Vision Computing*. New Zealand, 2012, pp. 43–48.

[D16] Becker D. *Private communication*. 2016.

[N11] Chapman H N et al. "Femtosecond X-ray protein nanocrystallography". In: *Nature* (2011): *Volume 470 issue 7332*, pp. 73–77.

[P16] *The Science Data Processor and Regional Centre Overview*. Cambridge, 12–13 April 2016.

[WG10] Eatough R P Molkenthin N Kramer M A Noutsos Keith M J Stappers B W and Lyne A G. "Selection of radio pulsar candidates using artificial neural networks". In: *Monthly Notices of the Royal Astronomical Society* (2010): *407*, pp. 2443–2450. DOI: 10.1111/j.1365-2966.2010.17082.x.

LSDMA-Driven Advances in Data Analysis

Thomas Jejkal[a], Bernd Schuller[b], Richard Grunzke[c]

[a] Karlsruhe Institute of Technology (KIT), Karlsruhe
[b] Forschungszentrum Jülich, Jülich Supercomputing Centre, Jülich
[c] Technische Universität Dresden, Dresden

Abstract Besides data management, the analysis of research data is another important aspect covered by LSDMA. The overarching goal of all data efforts must be to start managing the scientist's data right after the data left the data acquisition device as this is the only way to capture gapless provenance information. This is inevitable for transparent and reproducible science. However, this means also that the research data is at the very beginning of its lifecycle. In order to obtain publishable results the data often has to go through several processing steps until the final results are available. Such processing steps can reach from basic scripts to complex scientific workflows consisting of several dependent processing steps. In order to achieve the aforementioned reproducibility capturing the data provenance, e.g. what happened with the data and led to which results, should be an essential part of every processing step. This article describes the efforts taken by the LSDMA project in order to provide scientists with data analysis capabilities integrated seamlessly into data management workflows. For this, two approaches were chosen: The integration of data processing into an existing data repository system and the extension of an existing computing middleware by automated processing of data stored next to the computing environment. This article presents both approaches, realized use cases and finally gives a summary of the LSDMA efforts in term of data analysis.

1 Introduction

Management of large scale research data is challenging and it gets even more so when adding analysis workflows to the data. This is on the one hand due to the fact, that typical data management systems have no support for large

scale data analysis. On the other hand it gets more and more important to capture consistent provenance information in order to be able to reproduce analysis workflows. It is s no longer sufficient telling the user to download the data to its processing location and upload it after successful processing as the user often has no time or expertise to add a sufficient amount of provenance information. Therefor, a seamless integration of data processing environment and data management environment is highly desirable.
LSDMA Work Package 6 (WP6) mainly followed two approaches of integrating data management and data analysis: One approach aims at the integration of data driven workflows into UNICORE [97c; Ben16], which is primarily a middleware for computing tasks. Another approach deals with the integration of data processing as a service into the generic repository architecture KIT Data Manager [KIT16; Jej14]. Both approaches are presented in the following chapters. Finally, a potential integration of both solutions is discussed.

2 Integration of Data Processing into KIT Data Manager

A repository is a managed location in which collections of digital data objects are registered preserved, made accessible and retrievable, and are curated. It is essential that data in a digital data object is accompanied by metadata describing the data content and organization to enable their reuse in the future. Traditional repositories are holding digital data objects at their final lifecycle phase, where they have to be preserved and made accessible, e.g. for citation. In contrast, as research data repository, the KIT Data Manager mainly deals with data at the beginning of its lifecycle. Thus, after the ingest of a digital data object, data processing workflows have to be applied to the data in order to extract knowledge and new insights worth being published. Of course, even if the data is stored in a repository system, the user is still able to process the data outside of the repository, ingest the results again manually and add provenance information, e.g. input objects, applied algorithms and parameters, manually. The main disadvantage of this approach, apart from the need for manual steps, is the lack of reproducibility as it is very unlikely that the scientist takes care of capturing detailed provenance information for all processing steps applied to the raw data. Therefore, the challenge that has been addressed in LSDMA WP6 was to provide the seamless integration of both, data workflows and repository system with the following advantages:

- The repository system can take care of transferring data asynchronously in a proper representation to the execution environment.
- Once configured, data workflows can be executed and monitored by the repository system without user interaction.
- Workflow results can be ingested and linked to the input data object(s) automatically.
- The repository system can capture provenance information for all workflows.

In order to realize the goal of integrating data workflows and repository system, the repository system as well as suitable workflows have to fulfill a couple of requirements which are presented in the following sections. At the end of this chapter, some implemented use cases are presented and discussed.

2.1 Data Workflow Service

The basic ideas behind KIT Data Manager have already been presented in [Jej14]. Simply spoken, the data workflow service is an extension to the previously presented architecture introducing a new aspect to the internal metadata model and another high-level service. Both are following the principles of KIT Data Manager providing a high level of extensibility and flexibility allowing to exchange underlying technologies without affecting applications using the interfaces on top. Furthermore, existing structures and services are seamlessly integrated, e.g. the data workflow service metadata model was integrated into the existing metadata model of KIT Data Manager. The metadata model covers all information needed in order to describe, execute and monitor data workflows. The model basically consists of three major parts:

- **Data Workflow Task:** A data workflow task describes properties of one workflow step wrapping a generic user application. This includes application metadata, e.g. name, description, version and contact information, as well as all information needed for the actual execution, e.g. location of the application binary package, default arguments and mandatory environment properties. For the binary package a specific structure is pre-defined allowing the workflow service to execute all user applications in the same way.

- **Execution Environment:** The execution environment entity of the metadata model describes a processing environment capable of executing data workflow tasks. This includes basic metadata, e.g. name and description, but also environment specific elements like available environment properties, the location of the attached data storage and the according handler implementation used to execute tasks at the according execution environment. Details about the execution environment handler are described later.

- **Execution Environment Property:** Finally, there are execution environment properties that can be linked to execution environments to describe their capabilities and to data workflow tasks to describe their requirements. At execution time both environment properties lists can be compared to decide whether an execution environment is capable of executing a specific task or not. With this feature it is imaginable that an analysis workflow is executed in multiple configured execution environments if single workflow tasks have special requirements.

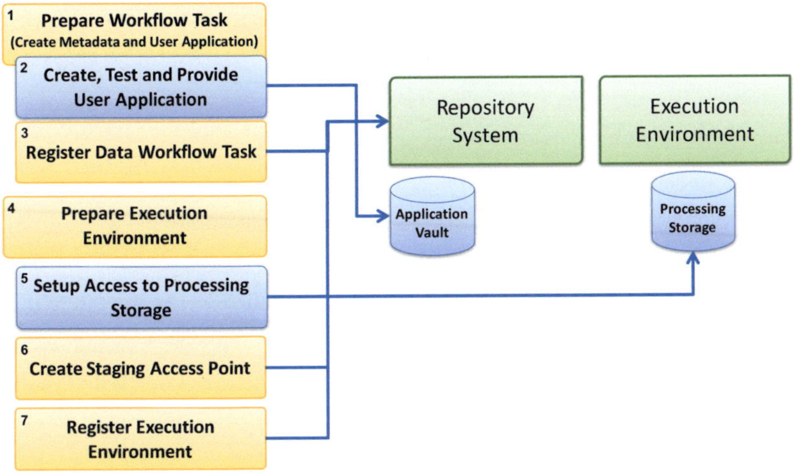

Figure 66: Registration procedure of a data workflow task and execution environment. Yellow boxes describe the creation of metadata entities, whereas blue boxes are detached steps that have to be performed by the application provider (step 2) or the operator of the repository system (step 5).

From the configuration perspective, registering data workflow tasks and execution environments is quite easy using the data workflow service of KIT Data Manager. This procedure is shown exemplarily in figure 66.

The most challenging part of the registration process is the preparation and testing of the user application. It has to be ensured that the user application, as soon as it is executed by the repository system, has a high reliability and a proper error handling. This necessitates extensive testing of the application and a detailed description of requirements towards the execution environment. This is mainly due to the fact that debugging an application run by the repository system is hardly possible. As soon as the application has been successfully tested and packaged it is stored in an application vault, which is a managed location accessible by the repository system. Once in the application vault an application package is never changed again. If the application changes it has to be registered as a new version of the same application. This is necessary in order to be able to reproduce previous executions with one specific version of the application. As one can see, the effort for preparing an application for the execution by the presented data workflow service is comparably high. That's the main reason why only recurring tasks should be implemented as data workflow tasks. However, by increasing the effort the quality of registered application is expected to be higher as double-checking the application avoids the need for later correction, redeployment and in the worst case reprocessing of data.

Another custom part of the data workflow preparation is the registration of the execution environment. Typically, this is done only once as the execution environment attached to a repository system, e.g. a compute cluster, does not change quite often. Basically, there are only two relevant properties of each execution environment: One is the configured access point allowing the repository system to access the processing storage, the other is a execution environment handler implementation taking care of the entire processing workflow including data staging, application preparation, application execution and monitoring as well as registering the result data as new digital data object in the repository system. Most of these steps are generic, e.g. data staging, application preparation and result registration. Only custom steps like the actual task submission and the monitoring are specific for each execution environment and must be implemented for new environments.

As soon as both data workflow task and execution environment are registered, the task can be assigned to a digital data object to run in the configured execution environment. The according process is shown and described in figure 67. Even if the setup and execution process looks complex, most of the effort has to be performed only once. Once configured, the overall benefit, e.g. automated processing of recurring tasks, automated ingest of results including gapless provenance information, greatly outweighs the effort. The following sections are presenting scientific use cases benefiting of the Data Workflow service.

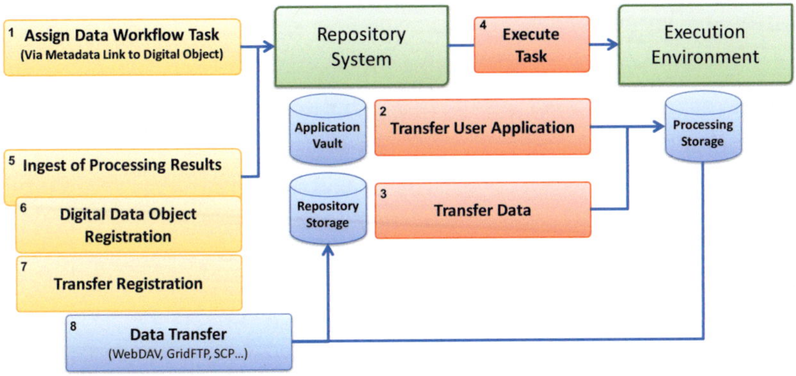

Figure 67: The figure presents the execution process of a data workflow. After assigning a task to a digital data object (1) the user application and the data are transferred to the configured processing storage (2, 3). Afterwards, the actual user application is executed by the repository system using the configured execution environment handler implementation (4). After successful execution, a new digital data object is created and the output data is ingested to the repository system (5, 6, 7, 8).

2.2 Scientific Use Cases

This section presents three scientific use cases integrating the data workflow service. From the data processing perspective, these use cases are quite simple as they allow data-parallel processing. Nevertheless, the automated processing, the reproducibility and the documentation of the data provenance are important advantages for the scientist independent from the complexity of the processing workflow.

Nanoscopy Open Reference Data Repository

The Nanoscopy Open Reference Data Repository (NORDR) is build up in Data Lifecycle Lab Key Technologies together with colleagues from Mannheim and Heidelberg. Datasets produced using lightoptical nanoscopy are stored automatically in a repository system [Pra15] and are to be processed using novel algorithms implemented in Matlab which are currently under development. The processing consists of three steps linked together using three different tasks executed by the data workflow service. The resulting provenance information is extracted into a ProvONE [Cue14] model and can be used to compare different workflows and the effect of changing particular input parameters or algorithm versions. By doing so, the development of the processing algorithms and the knowledge gain can be significantly advanced.

Ultra-Fast Synchrotron Tomography

The Ultra-Fast Synchrotron Tomography is a novel imaging method for studying moving biological samples [Yan13]. Due to a reduced number of projections the data acquisition time for living specimen can be essentially increased, but the effort for image reconstruction is considerably higher. Fortunately, the reconstruction process can be parallelized very well, such that volumes with 1024 slices can be processed in six minutes speeding up the processing time compared to the local execution by a factor of 120. Due to the fact, that the acquired data is ingested into a repository system right after the data acquisition, this processing can be realized by the repository system using the data workflow service. For parallel execution a Hadoop cluster was utilized via an appropriate execution environment handler implementation.

Software Workflow for the Automatic Tagging of Images

The Software Workflow for the Automatic Tagging of Images (SWATI) [Cha15a] provides algorithms for the automatic analysis of scans of medieval manuscripts. It allows to extract features from manuscript pages which can than be visualized and analyzed by codicology scholars. The main challenge in this project is the vast amount of digital data objects as each manuscript page stands for one processable object summing up to approx. 150.000 objects for one virtual collection. Each of these objects is processed in multiple steps resulting in an overall number of approx. 600.000 digital data objects after one processing cycle. The results are presented to experts using novel data visualization techniques. The tagged results then may lead to an evolution of processing algorithms which necessitates a reprocessing of the entire virtual collection. Due to the integration using the data workflow service this process can be fully automated allowing the scientists to concentrate on their research instead of data management.

3 Data-intensive Computing with UNICORE

UNICORE is a mature software for federating heterogeneous compute and data resources, comprising a full software stack from clients to various server components down to the components for accessing the actual compute or data resources. Its design follows several basic principles: abstraction of resource-specific details, openness and extensibility, operating system independence, security, and autonomy of resource providers. UNICORE is available from the SourceForge repository [97b] under a permissive, commercially friendly license (BSD).

3.1 General Developments regarding Data

In the course of LSDMA, the UNICORE middleware has seen a large number of improvements, as new functionality was added and new ways to use the software were explored. This development was driven by use cases and projects inside and outside of LSDMA, for example the European flagship Human Brain Project (HBP) [13b]. Within HBP, UNICORE is used as the middleware to integrate computing and data resources from leading European supercomputing centers, providing supercomputing capabilities and large-scale storage to neuroscientists.

A general overview of the current state of the software, its overall architecture and especially the crucial security features is beyond the scope of this work, but is presented in a recent paper [Ben16]. Here, we limit ourselves to describing UNICORE's features related to data-intensive computing, focusing on things that where added during the course of LSDMA.

UNICORE offers several interfaces for realizing data access and management, and is able to connect to a variety of backend systems that store the actual data and metadata.

- UNICORE's *Storages* are abstracted file system like data resources. They offer operations such as listing directories. To give access to files, storages act as factory services for file transfer resources. Different backends are supported. Beyond the traditional POSIX file systems, Amazon S3 [06a], Apache HDFS [Shv10] and CDMI [CDM13] can be used.

- UNICORE supports *metadata* [NS10]. The default implementation of the metadata manager stores the metadata in hidden files on the same storage as the data.

- *File transfers* are used to read from or write to a physical remote file. UNICORE supports both client-server data transfers and server-server transfers, with a number of available data transport protocols. For high-performance transfer of large volumes of data, the UNICORE File Transfer Protocol (UFTP) [SP11] can be used.

For data processing, UNICORE is based on the batch job model, where jobs are processed asynchronously by a backend cluster resource manager such as SLURM [JYG02]. Recently we have added support for running jobs on Apache YARN [Vav13], which opens up the area of "big data" processing, and its integration with traditional cluster computing.
To widen the range of users, it is now possible to access UNICORE via a Web portal [Pet13]. This is mainly intended for non-expert users, who need a simple way to access computing and data resources. The portal does not give access to all UNICORE features, but focuses on simple creation and management of jobs, handling data and running workflows.
To satisfy demands for the integration of more complex UNICORE functionality into custom applications and science gateways, we have added a com-

plete set of RESTful APIs [SRB14], that are continually expanded to cover all of UNICORE's rich feature set. Usage of the RESTful APIs is much simpler than the usage of UNICORE's SOAP Web Services APIs, and can be done in all popular languages, including Python. The RESTful APIs and SOAP APIs can be used in parallel in a fully consistent manner.

3.2 Development of Data Oriented Processing and its Application

The Data Oriented Processing (DOP) [SGG13] of UNICORE is a move from the traditional computing triggered by users to a model where processing is triggered by the data itself. This is highly beneficial for big data use cases, metadata extraction, compression, and preprocessing, which can be inefficient with traditional jobs. With DOP, rule files are associated to a directory. When data is incoming, the rules get evaluated and matching actions are triggered. An example is to have a rule automatically create and write MD5 checksums for every incoming PDF file. DOP can be compared to the iRODS [Raj10] rule engine, which in contrast is focused on low computing requirements while executing actions within rules. This is due to the fact that iRODS rules can only run on the file server itself. DOP scales to high computing requirements as the rule executions are only limited by the number of cores available on the HPC resource at hand. Furthermore, DOP rules are completely user controlled and running in the security context of a user, meaning with the login of the user. This makes DOP much more flexible than the iRODS mechanism as there, only administrators can create rules and have them run under the administration login. Also, DOP integrates seamlessly with backend storages mentioned before. The DOP feature was released with the UNICORE 7 release.
We applied DOP to integrate KNIME [Ber08] workflow executions with HPC resources [Gru16b]. The motivation is that varying user data sets and workflows exist in KNIME. The challenge is enable parallelization and utilize in a

Listing 2 "The central section of the UNICORE rule is shown which, among other things, governs how an action is defined and triggered."[Gru16b]

```
{
DirectoryScan: { IncludeDirs: ["."], "Interval": "30", },
Rules: [ {
Name: BioHPCMeasurements_2880_1,
Match: ".*.zip",
Action: { Type: BATCH, Job: {
  Name: knime_headless,
  Imports: [{ From: "file://${UC_FILE_PATH}",
              To: workflow.zip },],
  ApplicationName: knime,
  ApplicationVersion: 2.11.3_headless,
  Parameters: {
   l: "/lustre/ssd/grunzke/2880_1",
   k: { From: 0, To: 9, Step: 1 },
   n: "10",
   w: "../workflow.zip",
   t: "8",
   tr: "${UC_FILE_PATH}",},
  Resources: {
   Memory: 20664M, CPUs: 8,
   Runtime: 1h, Queue: haswell,
   CPUsPerNode: 8,}},},
}, ... ],}
```

HPC resources in a usable manner. We used KNIME for the creation and execution of workflows but it lacks a generic HPC integration. The UNICORE middleware is utilized to enable generic HPC access. The HPC integration is them evaluated via the image analysis use case at hand.

The use case involves analyzing microscopy data that is of fundamental importance in the life sciences. Data in gigabyte and terabyte range is created via confocal and light sheet microscopes. While the users are experts in chemistry, biology, genetics, and microscopy, they are rarely experts in using HPC resources too. Meaning, a simple way to distribute analysis jobs is needed. Our use case involves a preprocessing pipeline for E. coli grown in a Mother Machine that process raw movie data, often with many hundred movies and

arbitrary rotation. It is desired to rotate them horizontally, find growth channels and crop them, and save each channel in a specific format.
KNIME, The Open Analytics Platform, is a desktop application that can be used to analyze data with workflows in a flexible way. Workflow nodes, that perform tasks on data, can be used to from large and complex workflows. KNIME provides graphical ease-of-use and is widely used in academia and industry. It can flexibly integrate data and tools by various means to access, transform, analyze, mine, visualize, and deploy the data. Many pre-developed nodes are freely available.
We integrated KNIME with HPC resources using the DOP feature. A rule (see Listing 2) is configured for a server side directory that is mounted on the workstation of the user. When a user wants to execute the workflow on the HPC cluster she just exports the workflow using the built-in KNIME export method to the directory. The workflow then gets executed on the HPC system. This results in a data set on the cluster that is extended with the results. We extended DOP to support parameter sweeps and utilized it here spawn several workflow instances that each analyze a separate chunk of the data.
To evaluate the approach, we first measured and computed the mean processing time and the speed-up with varying numbers of threads per workflow instance, which yielded a balance at 8 threads. Then, the induced overhead per workflow execution was measured at about 27s. The overhead is due to the interval of 30s between rule evaluations and UNICORE internal overhead. It seems large until one considers workflow executions times in the range of hours to days. Then, measurements were performed with up to 1.76 TB in 7.488 M files, see Figure 68. Using 800 cores this resulted in a runtime of about 2 hours, which is 200x faster than the 17 days that were needed before on a 4 core workstation.
In conclusion we utilized the new UNICORE DOP feature to seamlessly and generically integrate KNIME with HPC resources. Further work includes fully automating an existing pre-processing pipeline from the microscopy to

HPC to a metadata-driven repository. Also, more data-intensive use cases are planned besides work to offload individual KNIME workflow nodes to HPC.

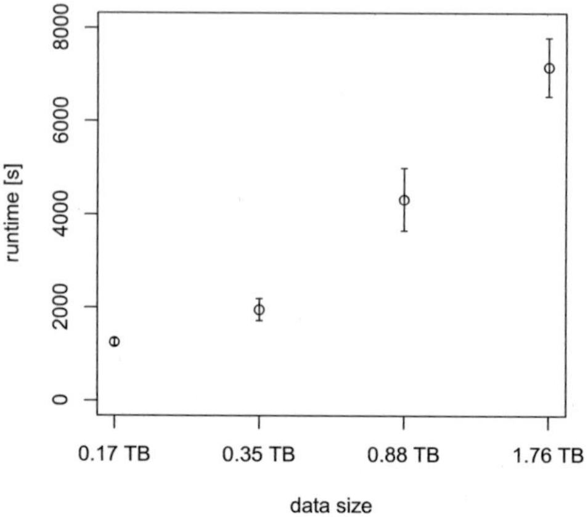

Figure 68: "Measurement results for concurrent processing of increasing number of datasets with an overall data size of up to 1.76 TB in 7.488 M files. The mean runtimes are 1259, 1955, 4318, and 7154 seconds respectively including error bars. These constitute a significant decrease in runtime compared to local processing on a workstation which would take about 17 days as compared to 2 hours on the HPC system."[Gru16b]

4 Summary

This article presented two approaches of integrating novel data analysis features into UNICORE and KIT Data Manager. In UNICORE the data oriented processing enables the scientist to schedule automated workflows by placing processing instructions next to the data. This allows intuitive usage on the one hand and provides application specific provenance information on the other hand. Provenance information was also a main aspect for the integration of the data workflow service into KIT Data Manager. There, analysis workflows are defined by metadata stored in a database allowing to reproduce the applied workflow tasks easily. Several scientific use cases have been imple-

mented and could prove the benefits of both solutions for the corresponding user community.
Despite of the independent development of both approaches the mutual integration of UNICORE's data oriented processing and KIT Data Manager's data workflow service are easily possible, e.g. by using UNICORE as execution environment for data workflow tasks.
For the future, the implementation of more use cases and improving both solutions with regard to scalability, security and usability are planned in order to establish flexible mechanisms for data analysis within the large scale data management landscape.

Quoted LSDMA Publications

[Ben16] Krzysztof Benedyczak et al. "UNICORE 7 - Middleware Services for Distributed and Federated Computing". In: *International Conference on High Performance Computing Simulation (HPCS)*. 2016. DOI: 10.1109/HPCSim.2016.7568392. URL: http://dx.doi.org/10.1109/HPCSim.2016.7568392.

[Cha15a] Swati Chandna et al. "Software workflow for the automatic tagging of medieval manuscript images (SWATI)". In: *SPIE/ IS&T Electronic Imaging*. International Society for Optics and Photonics. 2015, pp. 940206–940206.

[Gru16b] Richard Grunzke et al. "Metadata Management in the MoSGrid Science Gateway - Evaluation and the Expansion of Quantum Chemistry Support". In: *Journal of Grid Computing* (2016), pp. 1–13. ISSN: 1572-9184. DOI: 10.1007/s10723-016-9362-2. URL: http://dx.doi.org/10.1007/s10723-016-9362-2.

[Jej14] Thomas Jejkal et al. "KIT Data Manager: The Repository Architecture Enabling Cross-Disciplinary Research". In: *Large-Scale Data Management and Analysis (LSDMA) - Big Data in Science*. 2014, pp. 9–11. DOI: 10.5445/IR/1000043270.

[Pet13] Mariya Petrova et al. "The UNICORE Portal". In: *Proceedings of the 9th UNICORE Summit*. Vol. 21. 2013.

[Pra15] Ajinkya Prabhune et al. "An Optimized Generic Client Service API for Managing Large Datasets within a Data Repository". In: *Big Data Computing Service and Applications (BigDataService), 2015 IEEE First International Conference on*. Mar. 2015, pp. 44–51. DOI: 10.1109/BigDataService.2015.25.

[SGG13] Bernd Schuller, Richard Grunzke, and Andre Giesler. "Data Oriented Processing in UNICORE". In: *UNICORE Summit*

2013 Proceedings. Vol. 21. IAS Series. 2013, pp. 1–6. ISBN: 978-3-89336-910-2.

[SP11] Bernd Schuller and Tim Pohlmann. "UFTP: High-Performance Data Transfer for UNICORE". In: *Proceedings of the 7th UNICORE Summit*. IAS Series 9. Forschungszentrum Jülich GmbH, 2011, pp. 135–142.

[SRB14] Bernd Schuller, Jedrzej Rybicki, and Krzysztof Benedyczak. "High-Performance Computing on the Web: Extending UNICORE with RESTful Interfaces". In: *Proceedings of the Sixth International Conference on Advances in Future Internet*. IARIA XPS Press, 2014, pp. 35–38. ISBN: 978-1-61208-377-3. URL: http://www.thinkmind.org/%5C-index.php?view=article%5C&articleid=afin%5C_2014%5C_2%5C_10%5C_40020.

[Yan13] Xiaoli Yang et al. "Data Intensive Computing of X-Ray Computed Tomography Reconstruction at the LSDF". In: *Proceedings of the 21st Euromicro Intl. Conf. on Parallel, Distributed and Network-Based Computing (PDP'13)*. 2013. DOI: 10.1109/PDP.2013.21.

Other References

[06a] *Amazon Simple Storage Service (Amazon S3)*. Feb. 2006. URL: http://aws.amazon.com/s3/.

[13b] *Human Brain Project (HBP)*. Aug. 2013. URL: http://www.humanbrainproject.eu/.

[97b] *UNICORE Open Source project page*. Aug. 1997. URL: http://sourceforge.net/projects/unicore/.

[97c] *UNICORE Website*. Aug. 1997. URL: http://www.unicore.eu/.

[Ber08] Michael R Berthold et al. *KNIME: The Konstanz Information Miner*. Springer, 2008.

[CDM13] CDMI. *Cloud Data Management Interface (CDMI)*. Sept. 2013. URL: http://www.snia.org/tech%5C_activities/standards/curr%5C_standards/cdmi/.

[Cue14] V Cuevas-Vicentt et al. *ProvONE: A PROV Extension Data Model for Scientific Workflow Provenance*. Sept. 2014. URL: http://jenkins-1.dataone.org/jenkins/view/Documentation%5C%20Projects/job/ProvONE-Documentation-trunk/ws/provenance/ProvONE/v1/provone.html.

[JYG02] Morris A. Jette, Andy B. Yoo, and Mark Grondona. "SLURM: Simple Linux Utility for Resource Management". In: *In Lecture Notes in Computer Science: Proceedings of Job Scheduling Strategies for Parallel Processing (JSSPP) 2003*. Springer-Verlag, 2002, pp. 44–60.

[KIT16] KIT. *KIT Data Manager*. 2016. URL: http://datamanager.kit.edu.

[NS10] Waquas Noor and Bernd Schuller. "MMF: A flexible framework for metadata management in UNICORE". In: *Proceedings of the 6th UNICORE Summit*. Vol. 5. 2010, pp. 51–60. URL: http://hdl.handle.net/2128/3812.

[Raj10] Arcot Rajasekar et al. "iRODS primer: Integrated Rule-Oriented Data System". In: *Synthesis Lectures on Information Concepts, Retrieval, and Services* 2.1 (2010), pp. 1–143.

[Shv10] Konstantin Shvachko et al. "The Hadoop Distributed File System". In: *Proceedings of the 26th IEEE Symposium on Storage Systems and Technologies (MSST)*. 2010, pp. 1–10.

[Vav13] Vinod Kumar Vavilapalli et al. “Apache Hadoop Yarn: Yet Another Resource Negotiator”. In: *Proceedings of the 4th annual Symposium on Cloud Computing*. 2013, pp. 1–16.

Algorithm Engineering for Large Data Sets

Peter Sanders[a]

[a] Karlsruhe Institute of Technology (KIT), Karlsruhe

Abstract We give an overview on algorithms for large data sets developed in the group of Peter Sanders in the years 2012–2016. This includes algorithms for the (parallel) basic toolbox like sorting, searching, data structures, load balancing or communication efficient algorithms. Another focus is graph algorithms, in particular (hyper)graph-partitioning.

1 Introduction

Application data sets from various sources have grown much faster than the available computational resources which are still governed by Moore's law but increasingly hit physical limitations like clock frequency, energy consumption, and reliability. To name just a few applications, one can mention sensor data from particle colliders like LHC at CERN, the world wide web, sequenced genome data – ultimately from most human individuals, or GPS traces from millions and millions of smart phone users that can yield valuable information, e.g., on the current traffic situation. Large data sets are a fascinating topic for computer science in general and for algorithmics in particular. On the one hand, the applications can have enormous effects for our daily life, on the other hand, they are a big challenge for research and engineering. The main difficulty is that a successful solution has to take into account issues from three quite different areas of expertise: The particular application at hand, technological challenges, and the "traditional" areas of computer science know-how. This paper focuses on the aspect of algorithm design which is often at the core of underlying application problem. Integrating the three aspects means that we have to take into account traditional algorithmic know-how, technological aspects and the peculiarities of applications, e.g., latency requirements or properties of the input instances. Algorithm engineering [San09] with its emphasis on realistic models and its cycle of design, analysis, implementation, and experimental evaluation can serve as a glue between these requirements. Figure 69 summarizes the different aspects of algorithm engineering.

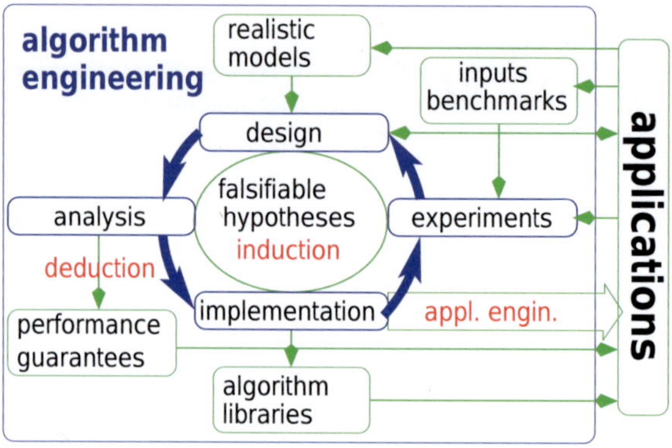

Figure 69: Algorithm Engineering.

This paper will give examples from the work of my group. Most of our work relates to generic techniques that can be used in many applications and ranges from theoretical considerations over experimental evaluation to reusable tools released as open source software. For more application specific work refer to Section 6.

2 The Basic Toolbox – Sorting, Searching and Data Structures

Surprisingly, even the most basic algorithms and data structures used in almost all applications as they are taught in basic algorithms courses (e.g. [MS08]) still allow a lot of innovations when it comes to large data sets and modern architectures. See [San15] for an a analysis why this is the case. We have worked a lot on sorting [FSS13; BS13; BES15; Axt15] and related problems [BFO13]. This includes the most scalable and robust parallel sorting algorithms currently available [Axt15] and the first serious work on parallel string sorting [BS13; BES15]. Further topics include priority queues [BKS15; RSD15], search trees [AS16] and the fastest concurrent hash tables currently available [MSD16].

3 The Parallel Basic Toolbox – Communication and Coordination

When it comes to parallel computing, the basic toolbox has to be extended with techniques for coordinating the processors. For the largest inputs, it often turns out that communication bandwith between the processors of a distributed memory machine is the bottleneck. We have therefore identified *Communication Efficient Algorithms* as a focus of our research [SSM13; HS16]. Once more, even the most fundamental problems turn out to allow a lot of improvements. In a first paper introducing the problem [SSM13], we consider duplicate detection, distributed Bloom filters, database join, and linear programming as examples. A subsequent paper [HS16] addresses various top-k selection problems.
Another long-standing focus of our work are load balancing and scheduling algorithms [SSS13; SS12b; MSS15].

4 Succinct Data Structures

To achieve good performance in Big Data applications, it is often mandatory to keep the relevant data in main memory. However, large amounts of main memory are expensive and on a particular machine, the maximal installable memory has a fixed upper bound. Hence, a lot of work has recently focused on the question whether one can compress data in such a way that one can still work with it. My group, mostly driven by my Postdocs Johannes Fischer (now Professor at University of Dortmund) and Simon Gog, have done a lot of work in that direction [Mül14; BF14; AF14; Bil13; FS13; GV16a; GV16b; Gog14]. Most notably, Simon Gog's succinct data structure library (SDSL) [Gog14] is quickly developing into a universally used tool for implementing such data structures. A prominent application area is information retrieval [Bil13; GV16a; GV16b]. Another example, are dictionary data structures where the keys need not actually be stored [AF14; Mül14]. This sounds quite abstract and theoretical. However, our research on that topic was driven by a cooperation with SAP – such data structures make up a significant part of the space consumption in the SAP HANA data base [MRF14].

5 Graph Algorithms

Graph algorithms are a major focus of work in our group. A long standing subject are algorithms for computing shortest paths. For an extensive survey on route planning refer to [Bas16]. We also developed the first parallel algorithm for multi-objective shortest paths [SM13; EKS14].
A more recent focus of our work was on graph partitioning [SS12a; OSS12; SS13; SSS12; SSS14; Bad14; GMS16; ASS15; SS16a]. See [Bul14] for an extensive overview. Graph partitioning splits a graph into pieces of about equal size such that few edges are cut. This is a central problem for processing large graphs since it allows to store the pieces on different nodes of a distributed system or on disk. Our graph partitioning tool KaHIP can handle

the largest available graphs and has record breaking quality on a wide spectrum of instances. In particular on graphs with small cuts like road networks or with very complicated structures like social networks, it outperforms the previously used systems by a wide margin.

We have started to generalize our techniques to hypergraph partitioning where edges can connect more than two nodes [Sch16]. Already now, our partitioner KaHyPar is both faster and better than most competing tool with the exception of the PaToH tool which is faster but computes lower quality solutions.

A key insight we gained from developing these systems is that good heuristics for the NP-complete graph partitioning problem require efficient algorithms for even more fundamental graph algorithms that have polynomial time solutions. For example, we developed a fast and scalable algorithm for approximate weighted matching [Bir13].

We are also using the graph partitioner as an ingredient in algorithms for other graph problems like independent sets [LSS15; Lam16; Dah16]. Furthermore, we have developed graph drawing algorithms using similar techniques as for partitioning [MNS15].

An important issue are also benchmarks and generators for large graphs that allow to evaluate graph algorithms for large instances [Bad13; Bad14; SS16b].

6 Applications

Since at the heart of many big data applications are algorithms that determine their performance, we are also directly involved in some of these applications – usually in cooperations with domain experts.

A typical example of the myriad of applications of graph partitioning in scientific computing is parallel fluid flow using the Lattice Boltzmann approach [Fie12]. An interesting (and quite common) challenge here was that the simulation code is actually optimized for working on regular grids rather than arbitrary graphs. The solution here is to work with a coarse grained graph whose nodes are regular grids of various resolution.

A cooperation with SAP mostly revolves around basic toolbox issues and their application in the core algorithms of a main memory column-store database [Fär12; Wil13; WDS13a; MRF14; Mül14; Mül15].

An interesting application from computational biology is the analysis of sequences of 3D microscopic images. For example, this is needed to track cells in embryonal development. In turn, this seemingly specialistic applications is one of the main tools for understanding what genomes mean. We believe that many of the resulting question can be cast as graph theoretical questions leading to scalable algorithms with high-quality results [Ste16b].

A show case application of big data is tracking particles in particle accelerators. For example, the CMS detector of the CERN-LHC accelerator can be viewed as a high speed camera yielding 40 million pictures a second which have to be analyzed in real time. We have shown how to parallelize an important part of this task on GPUs [Fun14] and are currently recasting the problem as a graph analysis problem with the hope to improve both speed and accuracy.

We have also adapted classical load balancing techniques to the requirements of massively parallel computations on isotopes [Dor16].

7 Conclusion and Contributions

Partially induced by the participation in LSDMA, the focus of work in my group has shifted to algorithms for large data sets and (even) stronger emphasis on practical impact inside and outside of Helmholtz research. For me this does not mean to give up theory and basic research but to bridges gaps between theory and practice. For LSDMA, this meant mostly two things. On the one hand, our expertise in basic algorithms in the basic toolbox and graph algorithms gives us the competence to work as a kind of consultant in application projects. Publications like [Fie12; Dor16; Fun14] fall in this category but are only the tip of the iceberg because often, despite being useful and interesting for both sides, the results do not always warrant a publication. On the other hand, we distill algorithms into widely useful tools like KaHiP, KaHyPar and SDSL. These are useful not only for Helmholtz but on a global scale and have high scientific impact witnessed by many publications and citations. We view the development and maintainance of such tools as a task that fits the Helmholtz mission (Key Technologies) very well since it requires a long term effort that is difficult to sustain with classical university research. A third kind of applied research is perhaps even more interesting: Identifying a grand challenge application problem that cannot be solved with existing algorithms and working on it in an interdisciplinary team with a major investment of person power. The SAP cooperation can be viewed as such an example which helped substantially improving SAP Hana and helped spawning the HANA Vora project. More closely to Helmholtz, the cooperation on 3D+t microscopy is such an effort that is just starting. We are in the process of initiating further such efforts.

Quoted LSDMA Publications

[AS16] Yaroslav Akhremtsev and Peter Sanders. "Fast Parallel Operations on Search Trees". In: *2016 IEEE 23rd International Conference on High Performance Computing (HiPC)*. Dec. 2016, pp. 291–300. DOI: 10.1109/HiPC.2016.042.

[ASS15] Yaroslav Akhremtsev, Peter Sanders, and Christian Schulz. "(Semi-)External Algorithms for Graph Partitioning and Clustering". In: *17th Workshop on Algorithm Engineering and Experiments (ALENEX)*. SIAM, 2015, pp. 33–43. DOI: 10.1137/1.9781611973754.4.

[Axt15] Michael Axtmann et al. "Practical Massively Parallel Sorting". In: *27th ACM Symposium on Parallelism in Algorithms and Architectures, (SPAA)*. 2015. DOI: 10.1145/2755573.2755595.

[Bas16] Hannah Bast et al. "Route Planning in Transportation Networks". In: *Algorithm Engineering*. Ed. by Lasse Kliemann and Peter Sanders. Vol. 9230. LNCS. Springer, 2016.

[BFO13] Timo Bingmann, Johannes Fischer, and Vitaly Osipov. "Inducing Suffix and Lcp Arrays in External Memory". In: *15th Workshop on Algorithm Engineering and Experiments, (ALENEX)*. SIAM, 2013, pp. 88–102. DOI: 10.1137/1.9781611972931.8. URL: http://dx.doi.org/10.1137/1.9781611972931.8.

[BKS15] Timo Bingmann, Thomas Keh, and Peter Sanders. "A bulk-parallel priority queue in external memory with STXXL". In: *14th Symposium on Experimental Algorithms (SEA)*. LNCS. Springer, 2015. DOI: 10.1007/978-3-319-20086-6_3.

[OSS12] Vitaly Osipov, Peter Sanders, and Christian Schulz. "Engineering Graph Partitioning Algorithms". In: *Experimental Algorithms: 11th International Symposium, SEA 2012, Bordeaux, France, June 7-9, 2012. Proceedings*. Ed. by Ralf Klasing.

Berlin, Heidelberg: Springer Berlin Heidelberg, 2012, pp. 18–26. ISBN: 978-3-642-30850-5. DOI: 10.1007/978-3-642-30850-5_3.

[Sch16] Sebastian Schlag et al. "k-way Hypergraph Partitioning via n-Level Recursive Bisection". In: *18th Workshop on Algorithm Engineering and Experiments (ALENEX)*. SIAM, 2016, pp. 53–67. DOI: 10.1137/1.9781611974317.5.

[SSM13] P. Sanders, S. Schlag, and I. Müller. "Communication efficient algorithms for fundamental big data problems". In: *2013 IEEE International Conference on Big Data*. Oct. 2013, pp. 15–23. DOI: 10.1109/BigData.2013.6691549.

[WDS13a] M. Weidner, J. Dees, and P. Sanders. "Fast OLAP query execution in main memory on large data in a cluster". In: *2013 IEEE International Conference on Big Data*. Oct. 2013, pp. 518–524. DOI: 10.1109/BigData.2013.6691616.

Other References

[AF14] Julian Arz and Johannes Fischer. "LZ-Compressed String Dictionaries". In: *Data Compression Conference (DCC)*. 2014, pp. 322–331. DOI: 10.1109/DCC.2014.36. URL: http://dx.doi.org/10.1109/DCC.2014.36.

[Bad13] David A. Bader et al., eds. *10th DIMACS Implementation Challenge – Graph Partitioning and Graph Clustering*. Vol. 588. Contemporary Mathematics. AMS, 2013. ISBN: 978-0-8218-9038-7, 978-0-8218-9869-7. DOI: 10.1090/conm/588.

[Bad14] David A. Bader et al. "Benchmarking for Graph Clustering and Partitioning". In: *Encyclopedia of Social Network Analysis and Mining*. Springer, 2014, pp. 73–82. DOI: 10.1007/978-1-4614-

6170-8_23. URL: http://dx.doi.org/10.1007/978-1-4614-6170-8_23.

[BES15] Timo Bingmann, Andreas Eberle, and Peter Sanders. "Engineering Parallel String Sorting". In: *Algorithmica* (2015), pp. 1–52. ISSN: 1432-0541. DOI: 10.1007/s00453-015-0071-1. URL: http://dx.doi.org/10.1007/s00453-015-0071-1.

[BF14] Kai Beskers and Johannes Fischer. "High-Order Entropy Compressed Bit Vectors with Rank/Select". In: *Algorithms* 7.4 (2014), pp. 608–620. DOI: 10.3390/a7040608. URL: http://dx.doi.org/10.3390/a7040608.

[Bil13] Philip Bille et al. "Sparse Suffix Tree Construction in Small Space". In: *40th International Colloquium on Automata, Languages, and Programming (ICALP)*. 2013, pp. 148–159. DOI: 10.1007/978-3-642-39206-1_13. URL: http://dx.doi.org/10.1007/978-3-642-39206-1_13.

[Bir13] Marcel Birn et al. "Efficient Parallel and External Matching". In: *Euro-Par 2013 Parallel Processing: 19th International Conference, Aachen, Germany, August 26-30, 2013. Proceedings.* Ed. by Felix Wolf, Bernd Mohr, and Dieter an Mey. Berlin, Heidelberg: Springer Berlin Heidelberg, 2013, pp. 659–670. ISBN: 978-3-642-40047-6. DOI: 10.1007/978-3-642-40047-6_66. URL: http://dx.doi.org/10.1007/978-3-642-40047-6_66.

[BS13] Timo Bingmann and Peter Sanders. "Parallel String Sample Sort". In: *Algorithms – ESA 2013: 21st Annual European Symposium, Sophia Antipolis, France, September 2-4, 2013. Proceedings.* Ed. by Hans L. Bodlaender and Giuseppe F. Italiano. Berlin, Heidelberg: Springer Berlin Heidelberg, 2013, pp. 169–180. ISBN: 978-3-642-40450-4. DOI: 10.1007/978-3-642-40450-4_15. URL: http://dx.doi.org/10.1007/978-3-642-40450-4_15.

[Bul14] Aydin Buluc et al. "Recent advances in graph partitioning". In: *Algorithm Engineering*. Ed. by Lasse Kliemann and Peter Sanders. LNCS. to appear. Springer, 2014.

[Dah16] Jakob Dahlum et al. "Accelerating Local Search for the Maximum Independent Set Problem". In: *15th Symposium on Experimental Algorithms, (SEA)*. Vol. 9685. LNCS. 2016, pp. 118–133. DOI: 10.1007/978-3-319-38851-9_9. URL: http://dx.doi.org/10.1007/978-3-319-38851-9_9.

[Dor16] Elizaveta Dorofeeva et al. "On a Dynamic Scheduling Algorithm for Massively Parallel Computations of Atomic Isotopes". In: *VII European Congress on Computational Methods in Applied Sciences and Engineering (ECCOMAS)*. 2016.

[EKS14] Stephan Erb, Moritz Kobitzsch, and Peter Sanders. "Parallel Bi-Objective Shortest Paths using Weight-Balanced B-Trees with Bulk Updates". In: *13th Symposium on Experimental Algorithms (SEA)*. Vol. 8504. LNCS. Springer, 2014. DOI: 10.1007/978-3-319-07959-2_10.

[Fär12] Franz Färber et al. "The SAP HANA Database–An Architecture Overview." In: *IEEE Data Eng. Bull.* 35.1 (2012), pp. 28–33.

[Fie12] J. Fietz et al. "Optimized Hybrid Parallel Lattice Boltzmann Fluid Flow Simulations on Complex Geometries". In: *18th Euro-Par*. Vol. 7484. LNCS. Springer, 2012, pp. 818–829. DOI: 10.1007/978-3-642-32820-6_81.

[FS13] Johannes Fischer and Peter Sanders, eds. *24th Symposium on Combinatorial Pattern Matching (CPM)*. Vol. 7922. LNCS. Springer, 2013. ISBN: 978-3-642-38904-7. DOI: 10.1007/978-3-642-38905-4. URL: http://dx.doi.org/10.1007/978-3-642-38905-4.

[FSS13] Patrick Flick, Peter Sanders, and Jochen Speck. "Malleable Sorting". In: *27th International Symposium on Parallel & Distributed Processing (IPDPS)*. IEEE. 2013, pp. 418–426. DOI: 10.1109/IPDPS.2013.90.

[Fun14] Daniel Funke et al. "Parallel track reconstruction in CMS using the cellular automaton approach". In: *Journal of Physics: Conference Series* 513.5 (2014), p. 052010.

[GMS16] Roland Glantz, Henning Meyerhenke, and Christian Schulz. "Tree-Based Coarsening and Partitioning of Complex Networks". In: *ACM Journal of Experimental Algorithmics* 21.1 (2016), 1.6:1–1.6:20. DOI: 10.1145/2851496. URL: http://doi.acm.org/10.1145/2851496.

[Gog14] Simon Gog et al. "From Theory to Practice: Plug and Play with Succinct Data Structures". In: *Symposium on Experimental Algorithms*. Springer. 2014, pp. 26–337.

[GV16a] Simon Gog and Rossano Venturini. "Fast and Compact Hamming Distance Index". In: *39th ACM Conference on Research and Development in Information Retrieval (SIGIR)*. 2016, pp. 285–294. DOI: 10.1145/2911451.2911523. URL: http://doi.acm.org/10.1145/2911451.2911523.

[GV16b] Simon Gog and Rossano Venturini. "Succinct Data Structures in Information Retrieval: Theory and Practice". In: *39th ACM Conference on Research and Development in Information Retrieval (SIGIR)*. 2016, pp. 1231–1233. DOI: 10.1145/2911451.2914802. URL: http://doi.acm.org/10.1145/2911451.2914802.

[HS16] Lorenz Hübschle-Schneider and Peter Sanders. "Communication Efficient Algorithms for Top-*k* Selection Problems". In: *30th IEEE International Parallel & Distributed Processing Symposium (IPDPS)*. 2016. DOI: 10.1109/IPDPS.2016.45.

[Lam16] Sebastian Lamm et al. “Finding Near-Optimal Independent Sets at Scale”. In: *18th Workshop on Algorithm Engineering and Experiments (ALENEX)*. SIAM, 2016, pp. 138–150. DOI: 10.1137/1.9781611974317.12.

[LSS15] Sebastian Lamm, Peter Sanders, and Christian Schulz. “Graph Partitioning for Independent Sets”. In: *14th Symposium on Experimental Algorithms (SEA)*. Vol. 9125. LNCS. Springer, 2015, pp. 68–81. DOI: 10.1007/978-3-319-20086-6_6. URL: http://dx.doi.org/10.1007/978-3-319-20086-6_6.

[MNS15] Henning Meyerhenke, Martin Nöllenburg, and Christian Schulz. “Drawing Large Graphs by Multilevel Maxent-Stress Optimization”. In: *23rd Symposium on Graph Drawing (GD)*. Vol. 9411. LNCS. Springer, 2015, pp. 30–43. DOI: 10.1007/978-3-319-27261-0_3. URL: http://dx.doi.org/10.1007/978-3-319-27261-0_3.

[MRF14] Ingo Müller, Cornelius Ratsch, and Franz Färber. “Adaptive String Dictionary Compression in In-Memory Column-Store Database Systems”. In: *17th Conference on Extending Database Technology (EDBT)*. 2014, pp. 283–294. DOI: 10.5441/002/edbt.2014.27. URL: http://dx.doi.org/10.5441/002/edbt.2014.27.

[MS08] K. Mehlhorn and P. Sanders. *Algorithms and Data Structures — The Basic Toolbox*. Springer, 2008. DOI: 10.1007/978-3-540-77978-0.

[MSD16] Tobias Maier, Peter Sanders, and Roman Dementiev. “Concurrent hash tables: fast *and* general?(!)” In: *21st ACM SIGPLAN Symposium on Principles and Practice of Parallel Programming (PPoPP)*. 2016. DOI: 10.1145/2851141.2851188.

[MSS15] Tobias Maier, Peter Sanders, and Jochen Speck. "Locality Aware DAG-Scheduling for LU-Decomposition". In: *29th IEEE International Parallel and Distributed Processing Symposium, (IPDPS)*. 2015, pp. 82–92. DOI: 10.1109/IPDPS.2015.85. URL: http://dx.doi.org/10.1109/IPDPS.2015.85.

[Mül14] Ingo Müller et al. "Retrieval and Perfect Hashing Using Fingerprinting". In: *13th Symposium on Experimental Algorithms (SEA)*. Vol. 8504. LNCS. Springer, 2014, pp. 138–149. DOI: 10.1007/978-3-319-07959-2_12.

[Mül15] Ingo Müller et al. "Cache-Efficient Aggregation: Hashing Is Sorting". In: *ACM SIGMOD International Conference on Management of Data*. 2015, pp. 1123–1136. DOI: 10.1145/2723372.2747644. URL: http://doi.acm.org/10.1145/2723372.2747644.

[RSD15] Hamza Rihani, Peter Sanders, and Roman Dementiev. "MultiQueues: Simple Relaxed Concurrent Priority Queues". In: *27th ACM Symposium on Parallelism in Algorithms and Architectures (SPAA)*. 2015, pp. 80–82. DOI: 10.1145/2755573.2755616. URL: http://doi.acm.org/10.1145/2755573.2755616.

[San09] P. Sanders. "Algorithm Engineering – An Attempt at a Definition". In: *Efficient Algorithms*. Vol. 5760. LNCS. Springer, 2009, pp. 321–340.

[San15] Peter Sanders. "Parallel Algorithms Reconsidered". In: *32nd International Symposium on Theoretical Aspects of Computer Science (STACS)*. Vol. 30. Dagstuhl, 2015, pp. 10–18. DOI: 10.4230/LIPIcs.STACS.2015.10.

[SM13] Peter Sanders and Lawrence Mandow. "Parallel Label-Setting Multi-Objective Shortest Path Search". In: *27th IEEE International Parallel & Distributed Processing Symposium*. 2013, pp. 215–224. DOI: 10.1109/IPDPS.2013.89.

[SS12a] Peter Sanders and Christian Schulz. "Distributed Evolutionary Graph Partitioning". In: *ALENEX 2012*. SIAM, 2012, pp. 16–29. DOI: 10.1137/1.9781611972924.2.

[SS12b] Peter Sanders and Jochen Speck. "Energy Efficient Frequency Scaling and Scheduling for Malleable Tasks". In: *18th Euro-Par*. Vol. 7484. LNCS. Springer, 2012, pp. 167–178. DOI: 10.1007/978-3-642-32820-6_18.

[SS13] Peter Sanders and Christian Schulz. "Think Locally, Act Globally: Highly Balanced Graph Partitioning". In: *12th Symposium on Experimental Algorithms (SEA)*. Vol. 7933. LNCS. Springer, 2013, pp. 164–175. DOI: 10.1007/978-3-642-38527-8_16.

[SS16a] Peter Sanders and Christian Schulz. "Advanced Multilevel Node Separator Algorithms". In: *15th Symposium on Experimental Algorithms (SEA)*. Vol. 9125. LNCS. 2016, pp. 118–133. DOI: 10.1007/978-3-319-38851-9_20.

[SS16b] Peter Sanders and Christian Schulz. "Scalable generation of scale-free graphs". In: *Inf. Process. Lett.* 116.7 (2016), pp. 489–491. DOI: 10.1016/j.ipl.2016.02.004. URL: http://dx.doi.org/10.1016/j.ipl.2016.02.004.

[SSS12] Ilya Safro, Peter Sanders, and Christian Schulz. "Advanced Coarsening Schemes for Graph Partitioning". In: *11th International Symposium on Experimental Algorithms (SEA)*. Vol. 7276. LNCS. Springer, 2012, pp. 369–380. DOI: 10.1007/978-3-642-30850-5_32.

[SSS13] Peter Sanders, Johannes Singler, and Rob van Stee. "Real-time integrated prefetching and caching". In: *J. Scheduling* 16.1 (2013), pp. 47–58. DOI: 10.1007/s10951-012-0301-1.

[SSS14] Ilya Safro, Peter Sanders, and Christian Schulz. "Advanced Coarsening Schemes for Graph Partitioning". In: *ACM Journal of Experimental Algorithmics* 19.1 (2014). DOI: 10.1145/2670338. URL: http://doi.acm.org/10.1145/2670338.

[Ste16b] Johannes Stegmaier et al. "Generating semi-synthetic validation benchmarks for embryomics". In: *13th IEEE International Symposium on Biomedical Imaging (ISBI)*. 2016, pp. 684–688. DOI: 10.1109/ISBI.2016.7493359. URL: http://dx.doi.org/10.1109/ISBI.2016.7493359.

[Wil13] Thomas Willhalm et al. "Vectorizing Database Column Scans with Complex Predicates." In: *ADMS@ VLDB*. 2013, pp. 1–12.

Non LSDMA Publications Quoted in this Book

[01a] *NDGF, Nordic Data Grid Facility*. 2001. URL: http://www.ndgf.org.

[01b] *SAML*. 2001. URL: https://de.wikipedia.org/wiki/Security%5C_Assertion%5C_Markup%5C_Language.

[03] *Lustre*. 2003. URL: http://lustre.org/.

[04] *Solr, Open Source search engine*. 2004. URL: https://lucene.apache.org/solr/.

[05] *GridFTP V2*. 2005. URL: https://www.ogf.org/documents/GFD.47.pdf.

[06a] *Amazon Simple Storage Service (Amazon S3)*. Feb. 2006. URL: http://aws.amazon.com/s3/.

[06b] *Buildout, Python Software Build System*. 2006. URL: http://www.buildout.org/en/latest/.

[07] *Apache Tika Website*. 2007. URL: http://tika.apache.org/.

[08] *PyWPS, Python implementation of the Web Processing Service standard from the Open Geospatial Consortium*. 2008. URL: http://pywps.org/.

[09] *SRM, The Storage Resource Manager Interface*. 2009. URL: https://sdm.lbl.gov/srm/.

[10a] *FAIR*. 2010. URL: https://www.gsi.de/en/researchaccelerators/fair.htm.

[10b] *NFS 4.1 / pNFS*. 2010. URL: http://www.pnfs.org/.

[11a] *Apache ODE Management API*. 2011. URL: http://ode.apache.org/management-api.html.

[11b] *Arango DB*. 2011. URL: https://www.arangodb.com/.

[11c] *Earth System Grid Federation*. 2011. URL: http://esgf.llnl.gov/.

[11d] *Jülich Dedicated GPU Environment*. 2011. URL: http://www.fz-juelich.de/ias/jsc/EN/Expertise/Supercomputers/JUDGE/JUDGE%5C_node.html.

[11e] *Localiation Microscopy — Swagger UI*. 2011. URL: http://datamanager.kit.edu/masi/localizationmicroscopy/swagger-ui/%5C#/.

[11f] *ownCloud*. 2011. URL: https://owncloud.org/.

[11g] *Pyramid, Python web framework*. 2011. URL: http://www.pylonsproject.org/.

[12] *xrootd*. 2012. URL: http://xrootd.org/.

[13a] *Apache ODE*. 2013. URL: http://ode.apache.org/.

[13b] *Human Brain Project (HBP)*. Aug. 2013. URL: http://www.humanbrainproject.eu/.

[13c] *PyCSW, Python implementation of an OGC catalog service for the web*. 2013. URL: http://pycsw.org/.

[13d] *RDA, Research Data Allicance*. 2013. URL: https://www.rd-alliance.org/node.

[14a] *Ansible, IT automation tool in Python*. 2014. URL: https://www.ansible.com/.

[14b] *Macaroons, authorization credentials that support decentralized verification*. 2014. URL: https://github.com/rescrv/libmacaroons.

[14c] *OpenID Connect*. 2014. URL: http://openid.net/connect/.

[15a] *Envriplus - Integrated Solutions for Environmental Research*. 2015. URL: http://www.envriplus.eu/.

[15b] *EUDAT - A Pan-european Data Infrastructure*. 2015. URL: https://www.eudat.eu/.

[15c] *Jülich Research on Exascale Cluster Architectures*. 2015. URL: http://www.fz-juelich.de/ias/jsc/EN/Expertise/Supercomputers/JURECA/JURECA%5C_node.html.

[15d] *Research Data Alliance (RDA) - Research Data Sharing Without Barriers*. 2015. URL: https://rd-alliance.org/.

[15e] *The OpenGIS Web Processing Service (WPS) Interface Standard*. 2015. URL: http://www.opengeospatial.org/standards/wps.

[15f] *W3C PROV Standard Overview*. 2015. URL: https://www.w3.org/TR/prov-overview/.

[16a] *AARC Blueprint Architecture*. June 29, 2016. URL: https://goo.gl/imZFch.

[16b] *Apache SOLR Open Source Enterprise Search Platform*. 2016. URL: http://lucene.apache.org/solr/.

[16c] *Authentication and Authorisation for Research and Collaboration*. June 29, 2016. URL: https://aarc-project.eu/.

[16e] *bwDataDiss project web site*. 2016. URL: http://www.alwr-bw.de/kooperationen/bwdatadiss/.

[16g] *CEPH*. 2016. URL: http://www.Ceph.com.

[16h] *Conda, Open Source package management system*. 2016. URL: http://conda.pydata.org/docs/.

[16i] *CRISP*. 2016. URL: http://www.crisp-fp7.eu/about-crisp/.

[16j] *Digital Curation Centre - Metadata Standards*. 2016. URL: http://www.dcc.ac.uk/resources/metadata-standards.

[16k] *DynaFed*. 2016. URL: http://information-technology.web.cern.ch/.

[16l] *EUDAT project web site*. http://eudat.eu/. [Online; accessed 10-October-2016]. 2016.

[16m] *European legislation on reuse of public sector information*. Oct. 2016. URL: https://ec.europa.eu/digital-single-market/en/european-legislation-reuse-public-sector-information/.

[16n] *FTS, File Transfer Service*. 2016. URL: http://information-technology.web.cern.ch/services/file-transfer.

[16o] *German research foundation "Safeguarding Good Scientific Practice"*. Oct. 2016. URL: http://www.dfg.de/en/research%5C_funding/principles%5C_dfg%5C_funding/good%5C_scientific%5C_practice/.

[16p] *INDIGO - DataCloud*. June 29, 2016. URL: https://www.indigo-datacloud.eu/.

[16q] *Localiation Microscopy Meta Data Scheme*. Mar. 2016. URL: http://datamanager.kit.edu/masi/localizationmicroscopy/2016-03/LocalizationMicroscopy.xsd.

[16r] *Localiation Microscopy — Nanoscopy METS Profile*. 2016. URL: http://datamanager.kit.edu/masi/localizationmicroscopy/mets/nanoscopy-METS-profile.xml.

[16s] *OGC Web Processing Service (WPS)*. 2016. URL: http://www.opengeospatial.org/standards/wps.

[16t] *Open Data repository*. Oct. 2016. URL: http://www.open-access.net/.

[16u] *RADAR project web site*. Oct. 2016. URL: http://www.radar-projekt.org/display/RE/Home/.

[16v] *Sony news release on magenetic tape technology*. Oct. 2016. URL: http://www.sony.net/SonyInfo/News/Press/201404/14-044E/.

[16w] *Thredds Data Server*. 2016. URL: http://www.unidata.ucar.edu/software/thredds/current/tds/TDS.html.

[16x] *Web processing service for climate impact and extreme weather event analyses, Nils Hempelmann*. 2016. URL: https://hal.archives-ouvertes.fr/hal-01375615v2.

[17a] *HPSS Collaboration web site*. http://www.hpss-collaboration.org/. [Online; accessed 29-March-2017]. 2017.

[17b] *Petra III, Synchrotron Radiation Source at DESY*. Deutsches Elektronen Synchrotron DESY. Notkestrasse 85, 22607 Hamburg, Germany, 2017. URL: http://photon-science.desy.de/facilities/petra%5C_iii/index%5C_eng.html.

[17c] *XFEL*. 2017. URL: http://www.xfel.eu/.

[67] *FERMIlab*. 1967. URL: http://www.fnal.gov.

[69] *GSI*. 1969. URL: https://www.gsi.de/en/start/news.htm.

[92] *HPSS, Hierarchical Parallel Storage System*. 1992. URL: http://www.hpss-collaboration.org/.

[93] *ALICE*. 1993. URL: https://home.cern/about/experiments/alice.

[97a] *SNIA, Storage Networking Industry Association*. 1997. URL: http://www.snia.org/.

[97b] *UNICORE Open Source project page*. Aug. 1997. URL: http://sourceforge.net/projects/unicore/.

[97c] *UNICORE Website*. Aug. 1997. URL: http://www.unicore.eu/.

[98] *GPFS, General Parallel File System*. 1998. URL: https://en.wikipedia.org/wiki/IBM%5C_General%5C_Parallel%5C_File%5C_System.

[99] *Apache Lucene Website*. 1999. URL: http://lucene.apache.org/.

[AB07] Nicola Armaroli and Vincenzo Balzani. "The future of energy supply: Challenges and opportunities". In: *Angewandte Chemie International Edition* 46.1-2 (2007), pp. 52–66. ISSN: 1521-3773. DOI: 10.1002/anie.200602373.

[AF14] Julian Arz and Johannes Fischer. "LZ-Compressed String Dictionaries". In: *Data Compression Conference (DCC)*. 2014, pp. 322–331. DOI: 10.1109/DCC.2014.36. URL: http://dx.doi.org/10.1109/DCC.2014.36.

[AHC14] Barty A Kirian R A Maia F R N C Hantke M Yoon C H White T A, Chapman H, and Cheetah. "Software for high-throughput reduction and analysis of serial femtosecond X-ray diffraction data". In: *Journal of Applied Crystallography* (2014): *47, 3*, pp. 1118–1131. DOI: 10.1107/S1600576714007626.

[al15] Dewdney P et al. *SKA1 system baseline v2 description. Rev 01*. Tech. rep. Science and Technology Facilities Council, 2015.

[al16] Ayyer K et al. "Macromolecular diffractive imaging using imperfect crystals". In: *Nature* (2016): *16949*.

[Asc16] K.P. Aschenbrenner et al. "Compressed sensing denoising for segmentation of localization microscopy data". In: *IEEE Int. Conf. Comput. Intelligence Bioinf. Comput. Biol. (CIBCB 2016, 5.-7.10.2016)* (16), in press.

[Axe11] M. et al. Axer. "A novel approach to the human connectome: Ultra-high resolution mapping of fiber tract in the brain". In: *NeuroImage* (2011): *vol.54, no. 2*, pp. 1091–1101.

[Bad13] David A. Bader et al., eds. *10th DIMACS Implementation Challenge – Graph Partitioning and Graph Clustering*. Vol. 588. Contemporary Mathematics. AMS, 2013. ISBN: 978-0-8218-9038-7, 978-0-8218-9869-7. DOI: 10.1090/conm/588.

[Bad14] David A. Bader et al. "Benchmarking for Graph Clustering and Partitioning". In: *Encyclopedia of Social Network Analysis and Mining*. Springer, 2014, pp. 73–82. DOI: 10.1007/978-1-4614-6170-8_23. URL: http://dx.doi.org/10.1007/978-1-4614-6170-8_23.

[Bar15] Christopher Bartz et al. "A Best Practice Analysis of HDF5 and NetCDF-4 Using Lustre". In: *High Performance Computing*. Ed. by Julian Martin Kunkel and Thomas Ludwig. Lecture Notes in Computer Science 9137. Frankfurt, Germany: Springer International Publishing, June 2015, pp. 274–281. ISBN: 978-3-319-20118-4. DOI: 10.1007/978-3-319-20119-1_20.

[Ber08] Michael R Berthold et al. *KNIME: The Konstanz Information Miner*. Springer, 2008.

[BES15] Timo Bingmann, Andreas Eberle, and Peter Sanders. "Engineering Parallel String Sorting". In: *Algorithmica* (2015), pp. 1–52. ISSN: 1432-0541. DOI: 10.1007/s00453-015-0071-1. URL: http://dx.doi.org/10.1007/s00453-015-0071-1.

[BF14] Kai Beskers and Johannes Fischer. "High-Order Entropy Compressed Bit Vectors with Rank/Select". In: *Algorithms* 7.4 (2014), pp. 608–620. DOI: 10.3390/a7040608. URL: http://dx.doi.org/10.3390/a7040608.

[Bil13] Philip Bille et al. "Sparse Suffix Tree Construction in Small Space". In: *40th International Colloquium on Automata, Languages, and Programming (ICALP)*. 2013, pp. 148–159. DOI:

10.1007/978-3-642-39206-1_13. URL: http://dx.doi.org/10.1007/978-3-642-39206-1_13.

[Bin16] T. Bingmann et al. "Thrill: High-Performance Algorithmic Distributed Batch Data Processing with C++". In: *ArXiv e-prints* (Aug. 2016). arXiv: 1608.05634 [cs.DC].

[Bir13] Marcel Birn et al. "Efficient Parallel and External Matching". In: *Euro-Par 2013 Parallel Processing: 19th International Conference, Aachen, Germany, August 26-30, 2013. Proceedings*. Ed. by Felix Wolf, Bernd Mohr, and Dieter an Mey. Berlin, Heidelberg: Springer Berlin Heidelberg, 2013, pp. 659–670. ISBN: 978-3-642-40047-6. DOI: 10.1007/978-3-642-40047-6_66. URL: http://dx.doi.org/10.1007/978-3-642-40047-6_66.

[Boh10] M. Bohn et al. "Localization microscopy reveals expression-dependent parameters of chromatin nanostructure". In: *Biophysical Journal* 99 (2010), pp. 1358–1367.

[BS13] Timo Bingmann and Peter Sanders. "Parallel String Sample Sort". In: *Algorithms – ESA 2013: 21st Annual European Symposium, Sophia Antipolis, France, September 2-4, 2013. Proceedings*. Ed. by Hans L. Bodlaender and Giuseppe F. Italiano. Berlin, Heidelberg: Springer Berlin Heidelberg, 2013, pp. 169–180. ISBN: 978-3-642-40450-4. DOI: 10.1007/978-3-642-40450-4_15. URL: http://dx.doi.org/10.1007/978-3-642-40450-4_15.

[Bul14] Aydin Buluc et al. "Recent advances in graph partitioning". In: *Algorithm Engineering*. Ed. by Lasse Kliemann and Peter Sanders. LNCS. to appear. Springer, 2014.

[Car02] L. Carl et al. *The Open Archives Initiative Protocol for Metadata Harvesting, Version 2.0 [Online]*. 2002. URL: http://www.openarchives.org/OAI/2.0/openarchivesprotocol.htm.

[CDM13] CDMI. *Cloud Data Management Interface (CDMI)*. Sept. 2013. URL: http://www.snia.org/tech%5C_activities/standards/curr%5C_standards/cdmi/.

[Cha14a] Konstantinos Chasapis et al. "Evaluating Lustre's Metadata Server on a Multi-socket Platform". In: *Proceedings of the 9th Parallel Data Storage Workshop*. PDSW 2014. New Orleans, Louisiana: IEEE Press, 2014, pp. 13–18. ISBN: 978-1-4799-7025-4. DOI: 10.1109/PDSW.2014.5.

[Cha15b] Konstantinos Chasapis et al. *Towards Scientific-Data Compression Using Variable Clustering*. Livermore, California, Aug. 2015.

[Coo13] John Cook et al. "Quantifying the consensus on anthropogenic global warming in the scientific literature". In: *Environmental Research Letters* 8.2 (2013), p. 024024. DOI: 10.1088/1748-9326/8/2/024024.

[CP12] Chen J P Spence J C and Millane R P. "Phase retrieval in femtosecond X–ray nanocrystallography". In: *Proceedings of the 27th Conference on Image and Vision Computing*. New Zealand, 2012, pp. 43–48.

[Cre11] C. Cremer et al. "Superresolution imaging of biological nanostructures by spectral precision distance microscopy". In: *Biotechnology Journal* 6 (2011), pp. 1037–1051.

[Cue14] V Cuevas-Vicentt et al. *ProvONE: A PROV Extension Data Model for Scientific Workflow Provenance*. Sept. 2014. URL: http://jenkins-1.dataone.org/jenkins/view/Documentation%5C%20Projects/job/ProvONE-Documentation-trunk/ws/provenance/ProvONE/v1/provone.html.

[Cue15] V. Cuevas-Vicenttín et al. “A PROV Extension Data Model for Scientific Workflow Provenance”. In: *Private communication* (2015).

[D14] Evanko D. “Taming the image background beast”. In: *Nature Methods* 228 (2014), p. 11.

[D16] Becker D. *Private communication.* 2016.

[Dah16] Jakob Dahlum et al. “Accelerating Local Search for the Maximum Independent Set Problem”. In: *15th Symposium on Experimental Algorithms, (SEA).* Vol. 9685. LNCS. 2016, pp. 118–133. DOI: 10.1007/978-3-319-38851-9_9. URL: http://dx.doi.org/10.1007/978-3-319-38851-9_9.

[Dar06] Sarah Darby. *The effectiveness of feedback on energy consumption.* Tech. rep. Environmental Change Institute, University of Oxford, 2006.

[Dem10] Bastian Demuth et al. “The UNICORE Rich Client: Facilitating the Automated Execution of Scientific Workflows”. In: *e-Science (e-Science), 2010 IEEE Sixth International Conference on.* IEEE. 2010, pp. 238–245.

[Dol15] Manuel F. Dolz et al. “An analytical methodology to derive power models based on hardware and software metrics”. In: *Computer Science - Research and Development* (2015), pp. 1–10. ISSN: 1865-2042. DOI: 10.1007/s00450-015-0298-8.

[Dom06] Ben Domenico et al. “Thematic Real-time Environmental Distributed Data Services (THREDDS): Incorporating Interactive Analysis Tools into NSDL”. In: *Journal of Digital Information* 2.4 (2006).

[Dor16] Elizaveta Dorofeeva et al. "On a Dynamic Scheduling Algorithm for Massively Parallel Computations of Atomic Isotopes". In: *VII European Congress on Computational Methods in Applied Sciences and Engineering (ECCOMAS)*. 2016.

[EKS14] Stephan Erb, Moritz Kobitzsch, and Peter Sanders. "Parallel Bi-Objective Shortest Paths using Weight-Balanced B-Trees with Bulk Updates". In: *13th Symposium on Experimental Algorithms (SEA)*. Vol. 8504. LNCS. Springer, 2014. DOI: 10.1007/978-3-319-07959-2_10.

[Fal14a] M. Falk et al. "Giving OMICS spatiotemporal dimensions by challenging microscopy: From functional networks to structural organization of cell nuclei elucidating mechanisms of complex radiation damage response and chromatin repair – PART A (Radiomics)". In: *Crit. Rev. Eukaryot. Gene Express.* 24 (2014), pp. 205–223.

[Fal14b] M. Falk et al. "Giving OMICS spatiotemporal dimensions by challenging microscopy: From functional networks to structural organization of cell nuclei elucidating mechanisms of complex radiation damage response and chromatin repair - PART B (Structuromics)". In: *Crit. Rev. Eukaryot. Gene Express.* 24 (2014), pp. 225–247.

[Fan13] Zhong Fan et al. "Smart grid communications: Overview of research challenges, solutions, and standardization activities". In: *Communications Surveys Tutorials, IEEE* 15.1 (2013), pp. 21–38. ISSN: 1553-877X. DOI: 10.1109/SURV.2011.122211.00021.

[Fär12] Franz Färber et al. "The SAP HANA Database–An Architecture Overview." In: *IEEE Data Eng. Bull.* 35.1 (2012), pp. 28–33.

[Fie12] J. Fietz et al. "Optimized Hybrid Parallel Lattice Boltzmann Fluid Flow Simulations on Complex Geometries". In: *18th Euro-Par*. Vol. 7484. LNCS. Springer, 2012, pp. 818–829. DOI: 10.1007/978-3-642-32820-6_81.

[FK10] English. In: *Pervasive Computing*. Ed. by Patrik Floréen and Jens Krüker. Vol. 6030. Lecture Notes in Computer Science. 2010, pp. 154–165. ISBN: 978-3-642-12653-6.

[FS13] Johannes Fischer and Peter Sanders, eds. *24th Symposium on Combinatorial Pattern Matching (CPM)*. Vol. 7922. LNCS. Springer, 2013. ISBN: 978-3-642-38904-7. DOI: 10.1007/978-3-642-38905-4. URL: http://dx.doi.org/10.1007/978-3-642-38905-4.

[FSS13] Patrick Flick, Peter Sanders, and Jochen Speck. "Malleable Sorting". In: *27th International Symposium on Parallel & Distributed Processing (IPDPS)*. IEEE. 2013, pp. 418–426. DOI: 10.1109/IPDPS.2013.90.

[Fun14] Daniel Funke et al. "Parallel track reconstruction in CMS using the cellular automaton approach". In: *Journal of Physics: Conference Series* 513.5 (2014), p. 052010.

[Gar11b] A.O. Garcia et al. *The Large Scale Data Facility: Data Intensive Computing for Scientific Experiments*. May 2011. DOI: 10.1109/IPDPS.2011.286.

[GJL12] Ulrich Greveler, Benjamin Justus, and Dennis Loehr. "Multimedia content identification through smart meter power usage profiles". In: *Computers, Privacy and Data Protection* 1 (2012), p. 10.

[GMS16] Roland Glantz, Henning Meyerhenke, and Christian Schulz. "Tree-Based Coarsening and Partitioning of Complex Networks". In: *ACM Journal of Experimental Algorithmics* 21.1

(2016), 1.6:1–1.6:20. DOI: 10.1145/2851496. URL: http://doi.acm.org/10.1145/2851496.

[Goe14] Christoph Goebel et al. “Energy Informatics”. In: *Business & Information Systems Engineering* 6.1 (2014), pp. 25–31. DOI: 10.1007/s12599-013-0304-2.

[Gog14] Simon Gog et al. “From Theory to Practice: Plug and Play with Succinct Data Structures”. In: *Symposium on Experimental Algorithms*. Springer. 2014, pp. 26–337.

[GRP10] Sidhant Gupta, Matthew S. Reynolds, and Shwetak N. Patel. “ElectriSense: single-point sensing using EMI for electrical event detection and classification in the home”. In: *Proceedings of the 12th ACM International Conference on Ubiquitous Computing*. UbiComp ’10. Copenhagen, Denmark: ACM, 2010, pp. 139–148. ISBN: 978-1-60558-843-8. DOI: 10.1145/1864349.1864375.

[GV16a] Simon Gog and Rossano Venturini. “Fast and Compact Hamming Distance Index”. In: *39th ACM Conference on Research and Development in Information Retrieval (SIGIR)*. 2016, pp. 285–294. DOI: 10.1145/2911451.2911523. URL: http://doi.acm.org/10.1145/2911451.2911523.

[GV16b] Simon Gog and Rossano Venturini. “Succinct Data Structures in Information Retrieval: Theory and Practice”. In: *39th ACM Conference on Research and Development in Information Retrieval (SIGIR)*. 2016, pp. 1231–1233. DOI: 10.1145/2911451.2914802. URL: http://doi.acm.org/10.1145/2911451.2914802.

[Har92] George William Hart. “Nonintrusive appliance load monitoring”. In: *Proceedings of the IEEE* 80.12 (1992), pp. 1870–1891.

[Hau16] M. Hausmann et al. "Challenges for super-resolution microscopy and fluorescent nano-probing: Understanding mechanisms behind tumour development and treatment". In: *Int J Cancer* (2016), submitted.

[Hic10] Jason Hick. "HPSS in the Extreme Scale Era: Report to DOE Office of Science on HPSS in 2018-2022". In: *Lawrence Berkeley National Laboratory* (2010).

[HS16] Lorenz Hübschle-Schneider and Peter Sanders. "Communication Efficient Algorithms for Top-*k* Selection Problems". In: *30th IEEE International Parallel & Distributed Processing Symposium (IPDPS)*. 2016. DOI: 10.1109/IPDPS.2016.45.

[HSM15] Lorenz Hübschle-Schneider, Peter Sanders, and Ingo Müller. "Communication Efficient Algorithms for Top-k Selection Problems". In: *CoRR* abs/1502.03942 (2015).

[Hua13] F. Huang et al. "Video-rate nanoscopy using sCMOS camera-specific single-molecule localization algorithms". In: *Nature Methods* 10 (7 2013), pp. 653–658.

[Joh99] Gage John. "Seurat's Silence". In: *Color and Meaning: Art, Science, and Symbolism*. University of California Press, 1999, pp. 223–225. ISBN: 9780520226111.

[JYG02] Morris A. Jette, Andy B. Yoo, and Mark Grondona. "SLURM: Simple Linux Utility for Resource Management". In: *In Lecture Notes in Computer Science: Proceedings of Job Scheduling Strategies for Parallel Processing (JSSPP) 2003*. Springer-Verlag, 2002, pp. 44–60.

[Kau11] R. Kaufmann et al. "Analysis of Her2/neu membrane protein cluster in different types of breast cancer cells using localization microscopy". In: *Journal of Microscopy* 242 (2011), pp. 46–54.

[KD12] G. Kalogridis and S. Dave. “PeHEMS: privacy enabled HEMS and load balancing prototype”. In: *Smart Grid Communications (SmartGridComm), 2012 IEEE Third International Conference on*. Nov. 2012, pp. 486–491. DOI: 10.1109/SmartGridComm.2012.6486032.

[KIT16] KIT. *KIT Data Manager*. 2016. URL: http://datamanager.kit.edu.

[KJ11] J Zico Kolter and Matthew J Johnson. “REDD: A public data set for energy disaggregation research”. In: *Workshop on Data Mining Applications in Sustainability (SIGKDD), San Diego, CA*. Vol. 25. Citeseer. 2011, pp. 59–62.

[KS14] Markus Karwe and Jens Strüker. “A survey on privacy in residential demand side management applications”. English. In: *Smart Grid Security*. Ed. by Jorge Cuellar. Vol. 8448. Lecture Notes in Computer Science. Springer International Publishing, 2014, pp. 154–165. ISBN: 978-3-319-10328-0. DOI: 10.1007/978-3-319-10329-7_10.

[Lam16] Sebastian Lamm et al. “Finding Near-Optimal Independent Sets at Scale”. In: *18th Workshop on Algorithm Engineering and Experiments (ALENEX)*. SIAM, 2016, pp. 138–150. DOI: 10.1137/1.9781611974317.12.

[Lan11] Betty Landesman. “Seeing Standards: A Visualization of the Metadata Universe”. In: *Technical Services Quarterly* 28.4 (2011), pp. 459–460. DOI: 10.1080/07317131.2011.598072.

[Lec13] Damien Lecarpentier et al. “EUDAT: A New Cross-Disciplinary Data Infrastructure for Science”. In: *International Journal of Digital Curation* 8.1 (2013), pp. 279–287.

[LSS15] Sebastian Lamm, Peter Sanders, and Christian Schulz. "Graph Partitioning for Independent Sets". In: *14th Symposium on Experimental Algorithms (SEA)*. Vol. 9125. LNCS. Springer, 2015, pp. 68–81. DOI: 10.1007/978-3-319-20086-6_6. URL: http://dx.doi.org/10.1007/978-3-319-20086-6_6.

[M10] Elad M. *Sparse and redundant representations: From theory to applications in signal and image processing, Haifa, Israel*. Springer Science+Business Media, New York, USA, 2010.

[Mar16] Valerio Mariani et al. "OnDA: online data analysis and feedback for serial X-ray imaging". In: *Journal of Applied Crystallography* 49.3 (June 2016), pp. 1073–1080. DOI: 10.1107/S1600576716007469.

[MAS15] MASi. *Metadata Management for Applied Sciences*. 2015. URL: http://www.scientific-metadata.de/.

[MBC13] Paolo Missier, Khalid Belhajjame, and James Cheney. "The W3C PROV Family of Specifications for Modelling Provenance Metadata". In: *Proceedings of the 16th International Conference on Extending Database Technology*. ACM. 2013, pp. 773–776.

[MNS15] Henning Meyerhenke, Martin Nöllenburg, and Christian Schulz. "Drawing Large Graphs by Multilevel Maxent-Stress Optimization". In: *23rd Symposium on Graph Drawing (GD)*. Vol. 9411. LNCS. Springer, 2015, pp. 30–43. DOI: 10.1007/978-3-319-27261-0_3. URL: http://dx.doi.org/10.1007/978-3-319-27261-0_3.

[Mol10] Andrés Molina-Markham et al. "Private memoirs of a smart meter". In: *Proceedings of the 2Nd ACM Workshop on Embedded Sensing Systems for Energy-Efficiency in Building*. BuildSys

'10. Zurich, Switzerland: ACM, 2010, pp. 61–66. ISBN: 978-1-4503-0458-0. DOI: 10.1145/1878431.1878446.

[Mor10] K.I. Mortensen et al. "Optimized localization analysis for single-molecule tracking and super-resolution microscopy". In: *Nature Methods* 7 (5 2010), pp. 377–381.

[MRF14] Ingo Müller, Cornelius Ratsch, and Franz Färber. "Adaptive String Dictionary Compression in In-Memory Column-Store Database Systems". In: *17th Conference on Extending Database Technology (EDBT)*. 2014, pp. 283–294. DOI: 10.5441/002/edbt.2014.27. URL: http://dx.doi.org/10.5441/002/edbt.2014.27.

[MS08] K. Mehlhorn and P. Sanders. *Algorithms and Data Structures — The Basic Toolbox*. Springer, 2008. DOI: 10.1007/978-3-540-77978-0.

[MSD16] Tobias Maier, Peter Sanders, and Roman Dementiev. "Concurrent hash tables: fast *and* general?(!)" In: *21st ACM SIGPLAN Symposium on Principles and Practice of Parallel Programming (PPoPP)*. 2016. DOI: 10.1145/2851141.2851188.

[MSS15] Tobias Maier, Peter Sanders, and Jochen Speck. "Locality Aware DAG-Scheduling for LU-Decomposition". In: *29th IEEE International Parallel and Distributed Processing Symposium, (IPDPS)*. 2015, pp. 82–92. DOI: 10.1109/IPDPS.2015.85. URL: http://dx.doi.org/10.1109/IPDPS.2015.85.

[Mül12] P. Müller et al. "Analysis of fluorescent nanostructures in biological systems by means of Spectral Position Determination Microscopy (SPDM)". In: *Current microscopy contributions to advances in science and technology*. Ed. by A. Méndez-Vilas. Vol. 1. 2012, pp. 3–12.

[Mül14] Ingo Müller et al. "Retrieval and Perfect Hashing Using Fingerprinting". In: *13th Symposium on Experimental Algorithms (SEA)*. Vol. 8504. LNCS. Springer, 2014, pp. 138–149. DOI: 10.1007/978-3-319-07959-2_12.

[Mül15] Ingo Müller et al. "Cache-Efficient Aggregation: Hashing Is Sorting". In: *ACM SIGMOD International Conference on Management of Data*. 2015, pp. 1123–1136. DOI: 10.1145/2723372.2747644. URL: http://doi.acm.org/10.1145/2723372.2747644.

[N11] Chapman H N et al. "Femtosecond X-ray protein nanocrystallography". In: *Nature* (2011): *Volume 470 issue 7332*, pp. 73–77.

[NS10] Waquas Noor and Bernd Schuller. "MMF: A flexible framework for metadata management in UNICORE". In: *Proceedings of the 6th UNICORE Summit*. Vol. 5. 2010, pp. 51–60. URL: http://hdl.handle.net/2128/3812.

[P16] *The Science Data Processor and Regional Centre Overview*. Cambridge, 12–13 April 2016.

[PW14] J. Potthof and J. van Wezel. "Landesprojekt bwDataArchiv – SCC erweitert digitales Archiv für Langzeitspeicherung von Forschungsdaten". In: *SCC News* 2014-1 (2014), pp. 12–14.

[Qua15] Volker Quaschning. *Regenerative Energiesysteme: Technologie – Berechnung – Simulation*. 9., aktualisierte und erw. Aufl. Hanser eLibrary. München: Hanser, 2015. ISBN: 978-3-446-44333-4. DOI: 10.3139/9783446443334.

[Raj10] Arcot Rajasekar et al. "iRODS primer: Integrated Rule-Oriented Data System". In: *Synthesis Lectures on Information Concepts, Retrieval, and Services* 2.1 (2010), pp. 1–143.

[RK13] David Reine and Mike Kahn. “Revisiting the Search for Long-Term Storage—A TCO Analysis of Tape and Disk”. In: *The Clipper Group Calculator, May* 13 (2013).

[RNS15] B. Rieger, R.P.J. Nieuwenhuizen, and S. Stallinga. “Image processing and Analysis for Single-Molecule Localization Microscopy: Computation for nanoscale imaging”. In: *IEEE signal Processing Magazine* 32 (2015), pp. 49–57.

[RSD15] Hamza Rihani, Peter Sanders, and Roman Dementiev. “MultiQueues: Simple Relaxed Concurrent Priority Queues”. In: *27th ACM Symposium on Parallelism in Algorithms and Architectures (SPAA)*. 2015, pp. 80–82. DOI: 10.1145/2755573.2755616. URL: http://doi.acm.org/10.1145/2755573.2755616.

[Sai03] P. Saiz et al. “AliEn – ALICE environment on the GRID”. In: *Nucl. Instrum. Meth.* A502 (2003), pp. 437–440. DOI: 10.1088/1742-6596/119/6/062012.

[San09] P. Sanders. “Algorithm Engineering – An Attempt at a Definition”. In: *Efficient Algorithms*. Vol. 5760. LNCS. Springer, 2009, pp. 321–340.

[San15] Peter Sanders. “Parallel Algorithms Reconsidered”. In: *32nd International Symposium on Theoretical Aspects of Computer Science (STACS)*. Vol. 30. Dagstuhl, 2015, pp. 10–18. DOI: 10.4230/LIPIcs.STACS.2015.10.

[Shi16] Jamie Shiers et al. *CERN Services for Long Term Data Preservation*. 2016.

[Shv10] Konstantin Shvachko et al. “The Hadoop Distributed File System”. In: *Proceedings of the 26th IEEE Symposium on Storage Systems and Technologies (MSST)*. 2010, pp. 1–10.

[SM13] Peter Sanders and Lawrence Mandow. "Parallel Label-Setting Multi-Objective Shortest Path Search". In: *27th IEEE International Parallel & Distributed Processing Symposium*. 2013, pp. 215–224. DOI: 10.1109/IPDPS.2013.89.

[SS12a] Peter Sanders and Christian Schulz. "Distributed Evolutionary Graph Partitioning". In: *ALENEX 2012*. SIAM, 2012, pp. 16–29. DOI: 10.1137/1.9781611972924.2.

[SS12b] Peter Sanders and Jochen Speck. "Energy Efficient Frequency Scaling and Scheduling for Malleable Tasks". In: *18th Euro-Par*. Vol. 7484. LNCS. Springer, 2012, pp. 167–178. DOI: 10.1007/978-3-642-32820-6_18.

[SS13] Peter Sanders and Christian Schulz. "Think Locally, Act Globally: Highly Balanced Graph Partitioning". In: *12th Symposium on Experimental Algorithms (SEA)*. Vol. 7933. LNCS. Springer, 2013, pp. 164–175. DOI: 10.1007/978-3-642-38527-8_16.

[SS14] A. Small and S. Stahlheber. "Fluorophore localization algorithms for super-resolution microscopy". In: *Nature Methods* 11 (3 2014), pp. 267–279.

[SS16a] Peter Sanders and Christian Schulz. "Advanced Multilevel Node Separator Algorithms". In: *15th Symposium on Experimental Algorithms (SEA)*. Vol. 9125. LNCS. 2016, pp. 118–133. DOI: 10.1007/978-3-319-38851-9_20.

[SS16b] Peter Sanders and Christian Schulz. "Scalable generation of scale-free graphs". In: *Inf. Process. Lett.* 116.7 (2016), pp. 489–491. DOI: 10.1016/j.ipl.2016.02.004. URL: http://dx.doi.org/10.1016/j.ipl.2016.02.004.

[SSS12] Ilya Safro, Peter Sanders, and Christian Schulz. "Advanced Coarsening Schemes for Graph Partitioning". In: *11th International Symposium on Experimental Algorithms (SEA)*.

Vol. 7276. LNCS. Springer, 2012, pp. 369–380. DOI: 10.1007/978-3-642-30850-5_32.

[SSS13] Peter Sanders, Johannes Singler, and Rob van Stee. "Real-time integrated prefetching and caching". In: *J. Scheduling* 16.1 (2013), pp. 47–58. DOI: 10.1007/s10951-012-0301-1.

[SSS14] Ilya Safro, Peter Sanders, and Christian Schulz. "Advanced Coarsening Schemes for Graph Partitioning". In: *ACM Journal of Experimental Algorithmics* 19.1 (2014). DOI: 10.1145/2670338. URL: http://doi.acm.org/10.1145/2670338.

[Sta10] R. Stallman. "Is digital inclusion a good thing? How can we make sure it is?" In: *Communications Magazine, IEEE* 48.2 (Feb. 2010), pp. 112–118. ISSN: 0163-6804. DOI: 10.1109/MCOM.2010.5402673.

[Sta16] Statista – Das Statistik-Portal. *Zuwachs der Weltbevölkerung, angegeben in unterschiedlichen Zeiteinheiten (Stand 2016)*. Online. 2016. URL: http://de.statista.com/statistik/daten/studie/1816/umfrage/zuwachs-der-weltbevoelkerung/.

[Ste16b] Johannes Stegmaier et al. "Generating semi-synthetic validation benchmarks for embryomics". In: *13th IEEE International Symposium on Biomedical Imaging (ISBI)*. 2016, pp. 684–688. DOI: 10.1109/ISBI.2016.7493359. URL: http://dx.doi.org/10.1109/ISBI.2016.7493359.

[Str10] A. Streit et al. "UNICORE 6 - Recent and Future Advancements". In: *Annals of Telecommunications-annales des Télécommunications* (2010), pp. 757–762.

[Tea16] DESY ASAP Team. *ASAP3: System Documentation*. System Documentation. Deutsches Elektronen Synchrotron DESY, 2016. URL: https://confluence.desy.de/display/ASAP3/ASAP3++Data+Storage+for+PETRA+III.

[TLW02] R.E. Thompson, D.R. Larson, and W.W. Webb. "Precise nanometer localization analysis for individual fluorescent probes". In: *Biophysical Journal* 82 (5 2002), pp. 2775–2783.

[Van04] Herbert Van de Sompel et al. "Resource Harvesting within the OAI-PMH Framework". In: *D-lib magazine* 10.12 (2004), pp. 1082–9873. URL: http://www.dlib.org/dlib/december04/vandesompel/12vandesompel.html.

[Vav13] Vinod Kumar Vavilapalli et al. "Apache Hadoop Yarn: Yet Another Resource Negotiator". In: *Proceedings of the 4th annual Symposium on Cloud Computing*. 2013, pp. 1–16.

[Wat05] Richard W Watson. "High performance storage system scalability: Architecture, implementation and experience". In: *22nd IEEE/13th NASA Goddard Conference on Mass Storage Systems and Technologies (MSST'05)*. IEEE. 2005, pp. 145–159.

[Wei98] Stuart Weibel et al. "Dublin Core Metadata for Resource Discovery". In: *Internet Engineering Task Force RFC* 2413 (1998), p. 222.

[WG10] Eatough R P Molkenthin N Kramer M A Noutsos Keith M J Stappers B W and Lyne A G. "Selection of radio pulsar candidates using artificial neural networks". In: *Monthly Notices of the Royal Astronomical Society* (2010): *407*, pp. 2443–2450. DOI: 10.1111/j.1365-2966.2010.17082.x.

[Wil13] Thomas Willhalm et al. "Vectorizing Database Column Scans with Complex Predicates." In: *ADMS@ VLDB*. 2013, pp. 1–12.

[Zha15] Y. Zhang et al. "Radiation induced chromatin conformation changes analysed by fluorescent localization microscopy, statistical physics, and graph theory". In: *PLoS ONE* 10 (2015), e0128555. DOI: 10.1371/journal.prone.0128555.

[ZR11] Michael Zeifman and Kurt Roth. “Nonintrusive appliance load monitoring: Review and outlook”. In: *Consumer Electronics, IEEE Transactions on* 57.1 (2011), pp. 76–84.

[ZWF06] Y. Zhao, M. Wilde, and I. Foster. “Applying the virtual data provenance model”. In: *Proceedings of the 2016 International Provenance and Annotation Workshop, Chicago USA* (2006), pp. 148–161.

List of all LSDMA Publications

[16d] *Birdhouse Open Source Project, Collection of WPS related components to support Climate data processing*. 2016. URL: http://bird-house.github.io/.

[16f] *Case Statement Research Data Repository Interoperability WG*. May 2016. URL: https://www.rd-alliance.org/group/research-data-repository-interoperability-wg/case-statement/research-data-repository.

[Agu15] Alvaro Aguilera et al. "Towards an Industry Data Gateway: An Integrated Platform for the Analysis of Wind Turbine Data". In: *Science Gateways (IWSG), 2015 7th International Workshop on*. June 2015, pp. 62–66. DOI: 10.1109/IWSG.2015.8.

[Ame14] Parinaz Ameri et al. "On the Application and Performance of MongoDB for Climate Satellite Data". In: *13th IEEE International Conference on Trust, Security and Privacy in Computing and Communications, TrustCom 2014, Beijing, China, September 24-26, 2014*. 2014, pp. 652–659. DOI: 10.1109/TrustCom.2014.84. URL: http://dx.doi.org/10.1109/TrustCom.2014.84.

[Ame16a] P. Ameri. "Chapter 6 - Database Techniques for Big Data". In: *Big Data*. Ed. by Rajkumar Buyya, Rodrigo N. Calheiros, and Amir Vahid Dastjerdi. Morgan Kaufmann, 2016, pp. 139–159. ISBN: 978-0-12-805394-2. DOI: 10.1016/B978-0-12-805394-2.00006-4. URL: http://www.sciencedirect.com/science/article/pii/B9780128053942000064.

[Ame16b] P. Ameri. “On a self-tuning index recommendation approach for databases”. In: *2016 IEEE 32nd International Conference on Data Engineering Workshops (ICDEW)*. May 2016, pp. 201–205. DOI: 10.1109/ICDEW.2016.7495648.

[Ame16c] Parinaz Ameri. “Challenges of Index Recommendation for Databases”. In: *Proceedings of the 28th GI-Workshop on Foundations of Databases (Grundlagen von Datenbanken),Nörten Hardenberg, Germany, May 24-27, 2016*. May 2016, pp. 10–14. URL: http://ceur-ws.org/Vol-1594/paper3.pdf.

[Ame16d] P. Ameri et al. “NoWog: A Workload Generator for Database Performance Benchmarking”. In: *2016 IEEE 14th Intl Conf on Dependable, Autonomic and Secure Computing, 14th Intl Conf on Pervasive Intelligence and Computing, 2nd Intl Conf on Big Data Intelligence and Computing and Cyber Science and Technology Congress (DASC/PiCom/DataCom/CyberSciTech)*. Aug. 2016, pp. 666–673. DOI: 10 . 1109 / DASC - PICom - DataCom-CyberSciTec.2016.120.

[AMS15] P. Ameri, J. Meyer, and A. Streit. “On a new approach to the index selection problem using mining algorithms”. In: *2015 IEEE International Conference on Big Data (Big Data)*. Oct. 2015, pp. 2801–2810. DOI: 10.1109/BigData.2015.7364084.

[Arg13] Lars Arge et al. “On (Dynamic) Range Minimum Queries in External Memory”. In: *Algorithms and Data Structures: 13th International Symposium, WADS 2013, London, ON, Canada, August 12-14, 2013. Proceedings*. Ed. by Frank Dehne, Roberto Solis-Oba, and Jörg-Rüdiger Sack. Berlin, Heidelberg: Springer Berlin Heidelberg, 2013, pp. 37–48. ISBN: 978-3-642-40104-6. DOI: 10.1007/978-3-642-40104-6_4. URL: http://dx.doi.org/10.1007/978-3-642-40104-6_4.

[AS15] Yaroslav Akhremtsev and Peter Sanders. "Fast Parallel Operations on Search Trees". In: *CoRR* abs/1510.05433 (2015). URL: http://arxiv.org/abs/1510.05433.

[AS16] Yaroslav Akhremtsev and Peter Sanders. "Fast Parallel Operations on Search Trees". In: *2016 IEEE 23rd International Conference on High Performance Computing (HiPC)*. Dec. 2016, pp. 291–300. DOI: 10.1109/HiPC.2016.042.

[ASS15] Yaroslav Akhremtsev, Peter Sanders, and Christian Schulz. "(Semi-)External Algorithms for Graph Partitioning and Clustering". In: *17th Workshop on Algorithm Engineering and Experiments (ALENEX)*. SIAM, 2015, pp. 33–43. DOI: 10.1137/1.9781611973754.4.

[Axt15] Michael Axtmann et al. "Practical Massively Parallel Sorting". In: *27th ACM Symposium on Parallelism in Algorithms and Architectures, (SPAA)*. 2015. DOI: 10.1145/2755573.2755595.

[Bas16] Hannah Bast et al. "Route Planning in Transportation Networks". In: *Algorithm Engineering*. Ed. by Lasse Kliemann and Peter Sanders. Vol. 9230. LNCS. Springer, 2016.

[Ben16] Krzysztof Benedyczak et al. "UNICORE 7 - Middleware Services for Distributed and Federated Computing". In: *International Conference on High Performance Computing Simulation (HPCS)*. 2016. DOI: 10.1109/HPCSim.2016.7568392. URL: http://dx.doi.org/10.1109/HPCSim.2016.7568392.

[Ber15] A. A. Bersenev et al. "An approach for integrating kerberized non web-based services with web-based identity federations". In: *Proceedings of the 10th International Conference on Software Paradigm Trends, ICSOFT 2015*. doi: 10.5220/0005509901440150. SCITEPRESS. 2015, pp. 144–150.

[BFO13] Timo Bingmann, Johannes Fischer, and Vitaly Osipov. "Inducing Suffix and Lcp Arrays in External Memory". In: *15th Workshop on Algorithm Engineering and Experiments, (ALENEX)*. SIAM, 2013, pp. 88–102. DOI: 10.1137/1.9781611972931.8. URL: http://dx.doi.org/10.1137/1.9781611972931.8.

[BKS15] Timo Bingmann, Thomas Keh, and Peter Sanders. "A bulk-parallel priority queue in external memory with STXXL". In: *14th Symposium on Experimental Algorithms (SEA)*. LNCS. Springer, 2015. DOI: 10.1007/978-3-319-20086-6_3.

[Bla15] Thomas Blank et al. "The Role of Energy Status Data in Solar Power Plants with Li-Ion Batteries". In: *Energy, Science and Technology 2015* (2015), p. 195.

[Ble12] M. Blessing et al. "Kilovoltage beam model for flat panel imaging system with bow-tie filter for scatter prediction and correction". In: *Physica Medica* 28.2 (2012), pp. 134–143. ISSN: 1120-1797. DOI: 10.1016/j.ejmp.2011.04.001. URL: //www.sciencedirect.com/science/article/pii/S1120179711000275.

[BS14] D. Becker and A. Streit. "A neural network based pre–selection of big data in photon science". In: BDCloud. 2014, pp. 71–76. DOI: 10.1109/BDCloud.2014.42.

[BS15a] D. Becker and A. Streit. *Localization of signal peaks in photon science imaging*. Tech. rep. UKSim, 2015. DOI: 10.1109/uksim.2015.35.

[BS15b] D. Becker and A. Streit. "Real–time signal identification in big data streams". In: *Bragg–spot localization in photon science*. 2015, pp. 611–616. DOI: 10.1109/HPCSim.2015.723710.

[BS16a] D Becker and A Streit. "Real-time Signal identification in Photon Science Imaging". In: *IJSSST* (2016).

[BS16b] D. Becker and A. Streit. "Realtime–Processing of Nanocrystallography Images". In: UKSim-AMSS. 2016. DOI: 10.1109/UKSim.2016.20.

[Bus16] Hannah Busch et al. "QuantiCod revisited. Neue Möglichkeiten zur Analyse mittelalterlicher Handschriften". In: *Book of Abstracts DHd 2015 " Von Daten zu Erkenntnissen"*. Graz, Feb. 2016. URL: http://gams.uni-graz.at/o:dhd2015.abstracts-gesamt.

[Cha14b] Konstantinos Chasapis et al. "Evaluating Power-Performace Benefits of Data Compression in HPC Storage Servers". In: *IARIA Conference*. Ed. by Steffen Fries and Petre Dini. Chamonix, France: IARIA XPS Press, Apr. 2014, pp. 29–34. ISBN: 978-1-61208-332-2.

[Cha15a] Swati Chandna et al. "Software workflow for the automatic tagging of medieval manuscript images (SWATI)". In: *SPIE/ IS&T Electronic Imaging*. International Society for Optics and Photonics. 2015, pp. 940206–940206.

[Deb12] M. Debatin et al. "CT reconstruction from few-views by Anisotropic Total Variation minimization". In: *2012 IEEE Nuclear Science Symposium and Medical Imaging Conference Record (NSS/MIC)*. Oct. 2012, pp. 2295–2296. DOI: 10.1109/NSSMIC.2012.6551521.

[Die15] Stefan Dietrich et al. *ASAP3: New Data Taking and Analysis Infrastructure for PETRA III*. HEPiX Workshop, Univ. Oxford/UK. Deutsches Elektronen Synchrotron DESY, 2015. URL: https://indico.cern.ch/event/346931/contributions/817807/attachments/684652/940445/Dietrich%5C_ASAP3%5C_new%5C_data%5C_taking%5C_infrastructure.pdf.

[Die16] Stefan Dietrich et al. *ASAP3: Status Update and Activities for XFEL*. HEPiX Workshop, DESY/Zeuthen. Deutsches Elektronen Synchrotron DESY, 2016. URL: https://indico.cern.ch/event/466991/contributions/1143592/attachments/1260614/1862916/Dietrich%5C_ASAP3%5C_Status%5C_Update%5C_and%5C_XFEL%5C_Activities.pdf.

[Emb14] Michael Embach et al. "eCodicology-Algorithms for the Automatic Tagging of Medieval Manuscripts". In: *The Linked TEI: Text Encoding in the Web* (2014), p. 172.

[End14] Florian Enders et al. *Nach der Digitalisierung. Zur computergestützten Erschließung mittelalterlicher Handschriften*. 2014. Chap. Nach der Digitalisierung. Zur computergestützten Erschließung mittelalterlicher Handschriften.

[Ert16] Benjamin Ertl et al. "Identity Harmonization for Federated HPC, Grid and Cloud Services". In: *Proceedings of the 2016 International Conference on High Performance Computing and Simulation*. IEEE. 2016, pp. 621–627.

[Gar11a] A.O. Garcia et al. "Data-intensive analysis for scientific experiments at the Large Scale Data Facility". In: *Large Data Analysis and Visualization (LDAV), 2011 IEEE Symposium on*. Oct. 2011, pp. 125–126. DOI: 10.1109/LDAV.2011.6092331.

[Ges12a] Sandra Gesing et al. "A Science Gateway Getting Ready for Serving the International Molecular Simulation Community". In: *Proceedings of Science* PoS(EGICF12-EMITC2)050 (Mar. 2012). URL: http://pos.sissa.it/archive/conferences/162/050/EGICF12-EMITC2%5C_050.pdf.

[Ges12b] Sandra Gesing et al. "A Single Sign-On Infrastructure for Science Gateways on a Use Case for Structural Bioinformatics". In: *Journal of Grid Computing* 10.4 (2012), pp. 769–790. ISSN:

1570-7873. DOI: 10.1007/s10723-012-9247-y. URL: http://link.springer.com/article/10.1007%2Fs10723-012-9247-y.

[Ges12c] Sandra Gesing et al. "A Single Sign-On Infrastructure for Science Gateways on a Use Case for Structural Bioinformatics". In: *Journal of Grid Computing* 10.4 (2012), pp. 769–790. ISSN: 1572-9184. DOI: 10.1007/s10723-012-9247-y.

[Ges12d] S. Gesing et al. "A Science Gateway Getting Ready for Serving the International Molecular Simulation Community". In: *EGI Community Forum 2012 / EMI Second Technical Conference*. Mar. 2012. URL: http://pos.sissa.it/archive/conferences/162/050/EGICF12-EMITC2_050.pdf.

[Ges14] Sandra Gesing et al. "Molecular Simulation Grid (MosGrid): A Science Gateway Tailored to the Molecular Simulation Community". English. In: *Science Gateways for Distributed Computing Infrastructures*. Springer International Publishing, 2014, pp. 151–165. ISBN: 978-3-319-11267-1. DOI: 10.1007/978-3-319-11268-8_11.

[Ges15a] Sandra Gesing et al. "Challenges and Modifications for Creating a MoSGrid Science Gateway for US and European Infrastructures". In: *Science Gateways (IWSG), 2015 7th International Workshop on*. June 2015, pp. 73–79. DOI: 10.1109/IWSG.2015.10.

[Ges15b] Sandra Gesing et al. "Science Gateways - Leveraging Modeling and Simulations in HPC Infrastructures via Increased Usability". In: *International Conference on High Performance Computing Simulation (HPCS)*. July 2015, pp. 19–26. DOI: 10.1109/HPCSim.2015.7237017.

[Gie17] Andre Giesler et al. "UniProv: A flexible Provenance Tracking System for UNICORE". In: *High-Performance Scien-*

tific Computing: First JARA-HPC Symposium, JHPCS 2016, Aachen, Germany, October 4–5, 2016, Revised Selected Papers. Springer International Publishing, 2017, pp. 233–242. ISBN: 978-3-319-53862-4. DOI: 10.1007/978-3-319-53862-4_20.

[GM14] Richard Grunzke and Ralph Müller-Pfefferkorn. "Certificate-free User-friendly HPC Access with UNICORE". In: *UNICORE Summit 2014 Proceedings*. Vol. 26. IAS Series. 2014, pp. 23–30. ISBN: 978-3-95806-004-3.

[Gru12] Richard Grunzke et al. "A Data Driven Science Gateway for Computational Workflows". In: *UNICORE Summit 2012 Proceedings*. Vol. 15. IAS Series. 2012, pp. 35–49. ISBN: 978–3-89336-829-7. URL: http://hdl.handle.net/2128/4705.

[Gru14a] Richard Grunzke et al. "Best Practices for Metadata Management in LSDMA". In: *Large-Scale Data Management and Analysis (LSDMA) - Big Data in Science*. Karlsruhe, 2014, pp. 32–33. DOI: 10.5445/IR/1000043270. URL: http://dx.doi.org/10.5445/IR/1000043270.

[Gru14b] Richard Grunzke et al. "Device-driven metadata management solutions for scientific big data use cases". In: *2014 22nd Euromicro International Conference on Parallel, Distributed, and Network-Based Processing*. IEEE. 2014, pp. 317–321.

[Gru14c] Richard Grunzke et al. "Improved Resilience and Usability for Science Gateway Infrastructures via Integrated Virtual Organizations". In: *EGI Community Forum 2014*. 2014.

[Gru14d] Richard Grunzke et al. "Standards-based Metadata Management for Molecular Simulations". In: *Concurrency and Computation: Practice and Experience* 26(10) (2014), pp. 1744–1759. ISSN: 1532-0634. DOI: 10.1002/cpe.3116.

[Gru14e] Richard Grunzke et al. "Towards Generic Metadata Management in Distributed Science Gateway Infrastructures". In: *IEEE/ACM CCGrid 2014 (14th International Symposium on Cluster, Cloud and Grid Computing)*. Chicago, IL, US, May 2014, pp. 566–570. DOI: 10.1109/CCGrid.2014.98.

[Gru15] Richard Grunzke et al. "Managing Complexity in Distributed Data Life Cycles Enhancing Scientific Discovery". In: *IEEE 11th International Conference on e-Science*. Aug. 2015, pp. 371–380. DOI: 10.1109/eScience.2015.72.

[Gru16a] Richard Grunzke. "Generic Metadata Handling in Scientific Data Life Cycles". PhD thesis. Doctoral Thesis, Technische Universität Dresden, Apr. 2016. URL: http://nbn-resolving.de/urn:nbn:de:bsz:14-qucosa-202070.

[Gru16b] Richard Grunzke et al. "Metadata Management in the MoSGrid Science Gateway - Evaluation and the Expansion of Quantum Chemistry Support". In: *Journal of Grid Computing* (2016), pp. 1–13. ISSN: 1572-9184. DOI: 10.1007/s10723-016-9362-2. URL: http://dx.doi.org/10.1007/s10723-016-9362-2.

[Gru16c] Richard Grunzke et al. "Towards a Metadata-driven Multi-community Research Data Management Service". In: *Proceedings of the 8th International Workshop on Science Gateways (IWSG 2016)*. Vol. 1871. CEUR-WS, 2016. URL: http://ceur-ws.org/Vol-1871/.

[H16] Heßling H. *Monte Carlo pathfinding in radio astronomy*. Tech. rep. GLOWSKA, 2016.

[Hag14] B. Hagemeier et al. "A Workflow for Polarized Light Imaging Using UNICORE Workflow Services". In: UNICORE Summit. Poznan, Poland, 2014.

[Har14] M. Hardt et al. "Combining the X.509 and the SAML Federated Identity Management Systems". In: *Proceedings of the 2nd International Conference, SNDS 2014 on Recent Trends in Computer Networks and Distributed Systems Security*. doi:10.1007/978-3-642-54525-2_36. Springer. 2014, pp. 404–415.

[Her12a] Sonja Herres-Pawlis et al. "Workflow-enhanced Conformational Analysis of Guanidine Zinc Complexes via a Science Gateway". In: *Studies in Health Technology and Informatics, 175:142-151, IOS Press*. 2012. DOI: 10.3233/978-1-61499-054-3-142.

[Her12b] Sonja Herres-Pawlis et al. "Workflow-enhanced conformational analysis of guanidine zinc complexes via a science gateway". In: *Volume 175: HealthGrid Applications and Technologies Meet Science Gateways for Life Sciences*. Studies in Health Technology and Informatics. 2012, pp. 142–151. DOI: 10.3233/978-1-61499-054-3-142.

[Her13a] Sonja Herres-Pawlis et al. "Orbital Analysis of Oxo and Peroxo Dicopper Complexes via Quantum Chemical Workflows in MoSGrid". In: *Proceedings of the International Workshop on Scientific Gateways 2013 (IWSG)*. 2013. URL: http://ceur-ws.org/Vol-993/paper3.pdf.

[Her13b] Sonja Herres-Pawlis et al. "User-Friendly Workflows in Quantum Chemistry". In: *Proceedings of the International Workshop on Scientific Gateways 2013 (IWSG)*. 2013. URL: http://ceur-ws.org/Vol-993/paper14.pdf.

[Her13c] S. Herres-Pawlis et al. "User-friendly metaworkflows in quantum chemistry". In: *2013 IEEE International Conference on Cluster Computing (CLUSTER)*. Sept. 2013, pp. 1–3. DOI: 10.1109/CLUSTER.2013.6702700.

[Her14a] Sonja Herres-Pawlis et al. "Expansion of Quantum Chemical Metadata for Workflows in the MoSGrid Science Gateway". In: *Science Gateways (IWSG), 2014 6th International Workshop on*. June 2014, pp. 67–72. DOI: 10.1109/IWSG.2014.18.

[Her14b] Sonja Herres-Pawlis et al. "Meta-Metaworkflows for Combining Quantum Chemistry and Molecular Dynamics in the MoSGrid Science Gateway". In: *6th International Workshop on Science Gateways (IWSG)*. June 2014, pp. 73–78. DOI: 10.1109/IWSG.2014.20.

[Her15a] Sonja Herres-Pawlis et al. "Multi-layer Meta-metaworkflows for the Evaluation of Solvent and Dispersion Effects in Transition Metal Systems Using the MoSGrid Science Gateways". In: *Science Gateways (IWSG), 2015 7th International Workshop on*. June 2015, pp. 47–52. DOI: 10.1109/IWSG.2015.13.

[Her15b] Sonja Herres-Pawlis et al. "Quantum Chemical Meta-Workflows in MoSGrid". In: *Concurrency and Computation: Practice and Experience* 27.2 (2015), pp. 344–357. ISSN: 1532-0634. DOI: 10.1002/cpe.3292.

[HGH14] Alexander Hoffmann, Richard Grunzke, and Sonja Herres-Pawlis. "Insights into the Influence of Dispersion Correction in the Theoretical Treatment of Guanidine-Quinoline Copper(I) Complexes". In: *Journal of Computational Chemistry* 35.27 (2014), pp. 1943–1950. ISSN: 1096-987X. DOI: 10.1002/jcc.23706.

[Hof13] Alexander Hoffmann et al. "User-friendly Metaworkflows in Quantum Chemistry". In: *IEEE International Conference on Cluster Computing (CLUSTER)*. IEEE. 2013, pp. 1–3. DOI: 10.1109/CLUSTER.2013.6702700.

[Jäk15] René Jäkel et al. "Architectural Implications for Exascale based on Big Data Workflow Requirements". English. In: *Big Data and High Performance Computing*. Vol. 26. Advances in Parallel Computing. IOS Press, 2015, pp. 101–113. ISBN: 978-1-61499-582-1. DOI: 10.3233/978-1-61499-583-8-101.

[Jej12] Thomas Jejkal et al. "LAMBDA – The LSDF Execution Framework for Data Intensive Applications". In: *Parallel, Distributed and Network-Based Processing (PDP), 20th Euromicro International Conference on*. submitted. 2012, pp. 213–220. DOI: 10.1109/PDP.2012.69. URL: http://dx.doi.org/10.1109/PDP.2012.69.

[Jej14] Thomas Jejkal et al. "KIT Data Manager: The Repository Architecture Enabling Cross-Disciplinary Research". In: *Large-Scale Data Management and Analysis (LSDMA) - Big Data in Science*. 2014, pp. 9–11. DOI: 10.5445/IR/1000043270.

[JS14] Christopher Jung and Achim Streit. *Large-Scale Data Management and Analysis (LSDMA) - Big Data in Science*. Springer, 2014. DOI: 10.5445/IR/1000043270.

[Jun14] C Jung et al. "Optimization of data life cycles". In: *Journal of Physics: Conference Series* 513.3 (2014), p. 032047. DOI: 10.1088/1742-6596/513/3/032047. URL: http://stacks.iop.org/1742-6596/513/i=3/a=032047.

[Jun15] C Jung et al. "Progress in Multi-Disciplinary Data Life Cycle Management". In: *Journal of Physics: Conference Series* 664.3 (2015), p. 032018. DOI: 10.1088/1742-6596/664/3/032018. URL: http://stacks.iop.org/1742-6596/664/i=3/a=032018.

[Kha12] Andranik Khachatryan et al. "Sensitivity of Self-tuning Histograms: Query Order Affecting Accuracy and Robustness". In: *Scientific and Statistical Database Management: 24th In-*

ternational Conference, SSDBM 2012, Chania, Crete, Greece, June 25-27, 2012. Proceedings. Ed. by Anastasia Ailamaki and Shawn Bowers. Berlin, Heidelberg: Springer Berlin Heidelberg, 2012, pp. 334–342. ISBN: 978-3-642-31235-9. DOI: 10.1007/978-3-642-31235-9_22.

[KKL14] Julian Kunkel, Michael Kuhn, and Thomas Ludwig. "Exascale Storage Systems – An Analytical Study of Expenses". In: *Supercomputing Frontiers and Innovations*. Volume 1, Number 1 (June 2014). Ed. by Jack Dongarra and Vladimir Voevodin, pp. 116–134. URL: http://superfri.org/superfri/article/view/20.

[KMB12] F. Keller, E. Muller, and K. Bohm. "HiCS: High Contrast Subspaces for Density-Based Outlier Ranking". In: *2012 IEEE 28th International Conference on Data Engineering*. Apr. 2012, pp. 1037–1048. DOI: 10.1109/ICDE.2012.88.

[KMe13] P. Kilpatrick, P. Milligan, and R. Stotzka (editors). "Proceedings of the 2013 21st Euromicro International Conference on Parallel, Distributed and Network-Based Processing (PDP 2013)". In: *2013 21st Euromicro International Conference on Parallel, Distributed, and Network-Based Processing*. Feb. 2013, pp. i–i. DOI: 10.1109/PDP.2013.1.

[Kob15] Andrei Y Kobitski et al. "An ensemble-averaged, cell density-based digital model of zebrafish embryo development derived from light-sheet microscopy data with single-cell resolution". In: *Scientific reports* 5 (2015).

[Krü14a] Jens Krüger et al. "Performance Studies on Distributed Virtual Screening". In: *BioMed Research International* (2014). DOI: 10.1155/2014/624024. URL: http://dx.doi.org/10.1155/2014/624024.

[Krü14b] Jens Krüger et al. "The MoSGrid Science Gateway - A Complete Solution for Molecular Simulations". In: *Journal of Chemical Theory and Computation* 10(6) (2014), pp. 2232–2245. DOI: 10.1021/ct500159h.

[Krü16] Jens Krüger et al. "Portals and Web-based Resources for Virtual Screening". In: *Current Drug Targets* 17 (2016), pp. 1–1. ISSN: 1389-4501/1873-5592. DOI: 10 . 2174 / 1389450117666160201105806. URL: http://www.eurekaselect.com/node/138922/article.

[KS16] Eileen Kuehn and Achim Streit. "Online Distance Measurement for Tree Data Event Streams." In: *DASC/PiCom/DataCom/CyberSciTech* (2016), pp. 681–688. DOI: 10.1109/DASC-PICom-DataCom-CyberSciTec.2016.122. URL: http://ieeexplore.ieee.org/document/7588920/.

[Kuh14] Michael Kuhn et al. "Compression By Default - Reducing Total Cost of Ownership of Storage Systems". In: *Supercomputing*. Ed. by Julian Martin Kunkel, Thomas Ludwig, and Hans Werner Meuer. Lecture Notes in Computer Science 8488. Leipzig, Germany: Springer International Publishing, June 2014. ISBN: 978-3-319-07517-4. DOI: 10.1007/978-3-319-07518-1.

[LAK15] Michael Lautenschlager, Panagiotis Adamidis, and Michael Kuhn. "Big Data Research at DKRZ – Climate Model Data Production Workflow". In: *Big Data and High Performance Computing*. 26th ed. Advances in Parallel Computing. IOS Press, 2015, pp. 133–155. ISBN: 978-1-61499-582-1. DOI: 10.3233/978-1-61499-583-8-133. URL: http://ebooks.iospress.nl/volume/big-data-and-high-performance-computing.

[Lut14] Richard Lutz et al. “Management of Meteorological Mass Data with MongoDB”. In: *28th International Conference on Informatics for Environmental Protection: ICT for Energy Effieciency, EnviroInfo 2014, Oldenburg, Germany, September 10-12, 2014.* 2014, pp. 549–556.

[Maa15] Ahmad Maatouki et al. “A Horizontally-Scalable Multiprocessing Platform Based on Node.js”. In: *2015 IEEE TrustCom/BigDataSE/ISPA, Helsinki, Finland, August 20-22, 2015, Volume 3.* 2015, pp. 100–107. DOI: 10.1109/Trustcom.2015.618. eprint: arXiv:1507.02798[cs.DC]. URL: http://dx.doi.org/10.1109/Trustcom.2015.618.

[McG15] Gary A. McGilvary et al. “Enhanced Usability of Managing Workflows in an Industrial Data Gateway”. In: *Interoperable Infrastructures for Interdisciplinary Big Data Sciences (IT4RIs 15).* Aug. 2015, pp. 495–502. DOI: 10.1109/eScience.2015.62.

[Mey14] Jörg Meyer et al. “Archival Services and Technologies for Scientific Data”. In: *Journal of Physics: Conference Series* 513.6 (2014), p. 062033. DOI: 10.1088/1742-6596/513/6/062033. URL: http://stacks.iop.org/1742-6596/513/i=6/a=062033.

[Mil14] Paul Millar et al. “Federated AAI: Enabling Collaboration”. In: *Big Data in Science, 1st edition.* 2014, pp. 22–24.

[Mil15] Millar et al. “Unlocking data: federated identity with LSDMA and dCache”. In: *Journal of Physics, Chep 2015.* doi:10.1088/1742-6596/664/4/042037. Conference Series by IOP Publishing, 2015.

[OSS12] Vitaly Osipov, Peter Sanders, and Christian Schulz. “Engineering Graph Partitioning Algorithms”. In: *Experimental Algorithms: 11th International Symposium, SEA 2012, Bordeaux, France, June 7-9, 2012. Proceedings.* Ed. by Ralf Klasing.

Berlin, Heidelberg: Springer Berlin Heidelberg, 2012, pp. 18–26. ISBN: 978-3-642-30850-5. DOI: 10.1007/978-3-642-30850-5_3.

[Ott13] JC Otte et al. "realTox: Real-time imaging of toxicant impact in whole organisms at single cell resolution". In: *NAUNYN-SCHMIEDEBERGS ARCHIVES OF PHARMACOLOGY*. Vol. 386. SPRINGER 233 SPRING ST, NEW YORK, NY 10013 USA. 2013, S60–S60.

[Pac13] Lars Packschies et al. "The MoSGrid e-Science Gateway: Molecular Simulations in a Distributed Computing Environment". In: *Journal of Cheminformatics* 5(Suppl 1):O3 (2013). DOI: 10.1186/1758-2946-5-S1-O3.

[Pet13] Mariya Petrova et al. "The UNICORE Portal". In: *Proceedings of the 9th UNICORE Summit*. Vol. 21. 2013.

[PHZ13] S. Pyatykh, J. Hesser, and L. Zheng. "Image Noise Level Estimation by Principal Component Analysis". In: *IEEE Transactions on Image Processing* 22.2 (Feb. 2013), pp. 687–699. ISSN: 1057-7149. DOI: 10.1109/TIP.2012.2221728.

[Pra15] Ajinkya Prabhune et al. "An Optimized Generic Client Service API for Managing Large Datasets within a Data Repository". In: *Big Data Computing Service and Applications (BigDataService), 2015 IEEE First International Conference on*. Mar. 2015, pp. 44–51. DOI: 10.1109/BigDataService.2015.25.

[Pra16a] Ajinkya Prabhune et al. "MetaStore: A Metadata Framework for Scientific Data Repositories". In: *2016 IEEE International Conference on Big Data*. IEEE. 2016, pp. 3026–3035.

[Pra16b] Ajinkya Prabhune et al. "Prov2ONE: An Algorithm for Automatically Constructing ProvONE Provenance Graphs". In: *Provenance and Annotation of Data and Processes: 6th Inter-*

national Provenance and Annotation Workshop, IPAW 2016, McLean, VA, USA, June 7-8, 2016, Proceedings. Springer International Publishing, 2016, pp. 204–208. ISBN: 978-3-319-40593-3. DOI: 10.1007/978-3-319-40593-3_22.

[PZH12] Stanislav Pyatykh, Lei Zheng, and Jürgen Hesser. "Efficient method of pixel neighborhood traversal". In: *Journal of Visual Communication and Image Representation* 23.5 (2012), pp. 719–728. ISSN: 1047-3203. DOI: 10.1016/j.jvcir.2012.03.008. URL: //www.sciencedirect.com/science/article/pii/S1047320312000582.

[PZH13] Stanislav Pyatykh, Lei Zheng, and Jürgen Hesser. *Fast noise variance estimation by principal component analysis*. 2013. DOI: 10.1117/12.2000276.

[Rig14] F. Rigoll et al. "A Privacy-Aware Architecture for Energy Management Systems in Smart Grids". In: *Ubiquitous Intelligence and Computing, 2014 IEEE 11th Intl Conf on and IEEE 11th Intl Conf on and Autonomic and Trusted Computing, and IEEE 14th Intl Conf on Scalable Computing and Communications and Its Associated Workshops (UTC-ATC-ScalCom)*. Dec. 2014, pp. 449–455. DOI: 10.1109/UIC-ATC-ScalCom.2014.9.

[Rig17] Fabian Rigoll. "Nutzerorientiertes Energiedatenmanagement". PhD thesis. Karlsruhe Institute of Technology, 2017. DOI: 10.5445/IR/1000068109.

[RS14] Fabian Rigoll and Hartmut Schmeck. "Konzeption eines Energiedatenmanagementsystems unter Beachtung von Datenschutz und Privatsphäre". In: *VDE-Kongress 2014*. Ed. by VDE. VDE. VDE VERLAG GmbH, Oct. 2014.

[Sch16] Sebastian Schlag et al. "*k*-way Hypergraph Partitioning via *n*-Level Recursive Bisection". In: *18th Workshop on Algorithm*

Engineering and Experiments (ALENEX). SIAM, 2016, pp. 53–67. DOI: 10.1137/1.9781611974317.5.

[SGG13] Bernd Schuller, Richard Grunzke, and Andre Giesler. "Data Oriented Processing in UNICORE". In: *UNICORE Summit 2013 Proceedings*. Vol. 21. IAS Series. 2013, pp. 1–6. ISBN: 978-3-89336-910-2.

[SP11] Bernd Schuller and Tim Pohlmann. "UFTP: High-Performance Data Transfer for UNICORE". In: *Proceedings of the 7th UNICORE Summit*. IAS Series 9. Forschungszentrum Jülich GmbH, 2011, pp. 135–142.

[SRB14] Bernd Schuller, Jedrzej Rybicki, and Krzysztof Benedyczak. "High-Performance Computing on the Web: Extending UNICORE with RESTful Interfaces". In: *Proceedings of the Sixth International Conference on Advances in Future Internet*. IARIA XPS Press, 2014, pp. 35–38. ISBN: 978-1-61208-377-3. URL: http://www.thinkmind.org/%5C-index.php?view=article%5C&articleid=afin%5C_2014%5C_2%5C_10%5C_40020.

[SSM13] P. Sanders, S. Schlag, and I. Müller. "Communication efficient algorithms for fundamental big data problems". In: *2013 IEEE International Conference on Big Data*. Oct. 2013, pp. 15–23. DOI: 10.1109/BigData.2013.6691549.

[Ste16a] Johannes Stegmaier et al. "Automation strategies for large-scale 3D image analysis". In: *at-Automatisierungstechnik* 64.7 (2016), pp. 555–566.

[Sto12] M. Stockhause et al. "Quality assessment concept of the World Data Center for Climate and its application to CMIP5 data". In: *Geoscientific Model Development* 5.4 (2012), pp. 1023–1032.

DOI: 10.5194/gmd-5-1023-2012. URL: http://www.geosci-model-dev.net/5/1023/2012/.

[Str15] M. Strutz et al. "ASAP3 - New Data Taking and Analysis Infrastructure for PETRA III". In: *J. Phys. Conf. Ser.* 664.4 (2015), p. 042053. DOI: 10.1088/1742-6596/664/4/042053.

[Sts12] D Stsepankou et al. "Evaluation of robustness of maximum likelihood cone-beam CT reconstruction with total variation regularization". In: *Physics in Medicine and Biology* 57.19 (2012), p. 5955. DOI: 10.1088/0031-9155/57/19/5955. URL: http://stacks.iop.org/0031-9155/57/i=19/a=5955.

[Sut12] M. Sutter et al. "File Systems and Access Technologies for the Large Scale Data Facility". In: *Remote Instrumentation for eScience and Related Aspects*. Ed. by Franco Davoli et al. New York, NY: Springer New York, 2012, pp. 239–256. ISBN: 978-1-4614-0508-5. DOI: 10.1007/978-1-4614-0508-5_16.

[SZ12] Nodari Sitchinava and Norbert Zeh. "A Parallel Buffer Tree". In: *Proceedings of the Twenty-fourth Annual ACM Symposium on Parallelism in Algorithms and Architectures*. SPAA '12. Pittsburgh, Pennsylvania, USA: ACM, 2012, pp. 214–223. ISBN: 978-1-4503-1213-4. DOI: 10.1145/2312005.2312046.

[Szu16] M. Szuba et al. "A Distributed System for Storing and Processing Data from Earth-Observing Satellites: System Design and Performance Evaluation of the Visualisation Tool". In: *2016 16th IEEE/ACM International Symposium on Cluster, Cloud and Grid Computing (CCGrid)*. May 2016, pp. 169–174. DOI: 10.1109/CCGrid.2016.19. eprint: arXiv:1511.07693[cs.DC].

[Teu12] Tanja Teuber et al. "Denoising by second order statistics". In: *Signal Processing* 92.12 (2012), pp. 2837–2847. ISSN: 0165-

1684. DOI: 10.1016/j.sigpro.2012.04.015. URL: http://www.sciencedirect.com/science/article/pii/S0165168412001375.

[Ton12] Danah Tonne et al. “A federated data zone for the arts and humanities”. In: *2012 20th Euromicro International Conference on Parallel, Distributed and Network-based Processing*. IEEE. 2012, pp. 198–205.

[Ton13] Danah Tonne et al. “Access to the DARIAH bit preservation service for humanities research data”. In: *2013 21st Euromicro International Conference on Parallel, Distributed, and Network-Based Processing*. IEEE. 2013, pp. 9–15.

[VRT12] Philipp Vanscheidt, Andrea Rapp, and Danah Tonne. “Storage Infrastructure of the Virtual Scriptorium St. Matthias”. In: *Digital Humanities 2012* (2012), p. 529.

[WDS13a] M. Weidner, J. Dees, and P. Sanders. “Fast OLAP query execution in main memory on large data in a cluster”. In: *2013 IEEE International Conference on Big Data*. Oct. 2013, pp. 518–524. DOI: 10.1109/BigData.2013.6691616.

[WDS13b] M. Weidner, J. Dees, and P. Sanders. “Fast OLAP query execution in main memory on large data in a cluster”. In: *2013 IEEE International Conference on Big Data*. Oct. 2013, pp. 518–524. DOI: 10.1109/BigData.2013.6691616.

[Wez15] J. van Wezel et al. “Towards an Interoperable Data Archive”. In: *Proceedings for the PV 2015 Conference*. Nov. 2015. URL: https://www.eumetsat.int/website/home/News/ConferencesandEvents/DAT_2447480.html.

[Yan13] Xiaoli Yang et al. “Data Intensive Computing of X-Ray Computed Tomography Reconstruction at the LSDF”. In: *Proceedings of the 21st Euromicro Intl. Conf. on Parallel, Distributed and Network-Based Computing (PDP'13)*. 2013. DOI: 10.1109/PDP.2013.21.

[Yan16] Xiaoli Yang. "Precise and Automated Tomographic Reconstruction with a Limited Number of Projections". PhD thesis. Karlsruhe, Karlsruher Institut für Technologie (KIT), Diss., 2016, 2016.

[ZGK17] Lukas Zimmermann, Richard Grunzke, and Jens Krüger. "Maintaining a Science Gateway - Lessons Learned from MoS-Grid". In: *Proceedings of the 50th Hawaii International Conference on System Sciences*. 2017. URL: http://hdl.handle.net/10125/41918.